D1289972

Guide to Teaching Computer Science

Orit Hazzan • Tami Lapidot • Noa Ragonis

Guide to Teaching Computer Science

An Activity-Based Approach

Second Edition

 Springer

Orit Hazzan
Dept. Education in Science & Technology
Technion—Israel Institute of Technology
Technion City
Haifa
Israel

Noa Ragonis
Computer Science Studies, Faculty
of Education
Beit Berl College;
Dept. Education in Science & Technology
Technion—Israel Institute of Technology
Doar Beit Berl
Israel

Tami Lapidot
Dept. Education in Science & Technology
Technion—Israel Institute of Technology
Technion City
Haifa
Israel

ISBN 978-1-4471-6629-0   ISBN 978-1-4471-6630-6 (eBook)
DOI 10.1007/978-1-4471-6630-6

Library of Congress Control Number: 2014956392

Springer London Heidelberg New York Dordrecht

Printed on acid-free paper

Springer is part of Springer Science+Business Media (www.springer.com)

To our families, students and colleagues

Prologue

This *Guide to Teaching Computer Science* can serve all computer science educators, both in high school and in academia, i.e., computer science university instructors, high school computer science teachers, and instructors of computer science teacher preparation programs. Specifically, the guide can be used as the textbook of the Methods of Teaching Computer Science (MTCS) course, offered to prospective and in-service computer science teachers. In all cases, the guide is organized in a way that enables an immediate application of its main ideas. This goal is achieved by presenting the rationale for addressing a variety of computer science education topics, as well as their detailed actual teaching process (including activities, worksheets, topics for discussions, and more).

The guide encompasses the authors' teaching and research experience in computer science education gained during the past three decades. Specifically, we have taught courses on computer science and computer science education to high school computer science pupils, undergraduate computer science students, and pre-service and in-service computer science teachers. In parallel, we have conducted research on a variety of computer science education topics, such as teaching methods, learning processes, teacher preparation, and social issues of computer science education.

In the second edition, we updated all the chapters with both content and references, and added 15 new activities; specifically, we highlighted current teaching approaches and trends to be integrated in the MTCS course.

We would like to thank all who contributed to our understanding of the nature of computer science education and fostered the approach presented in this guide: our students in the MTCS courses, high school classes, and in-service high school teacher professional development programs, as well as colleagues, researchers and instructors who collaborated with us in a variety of research and development projects. During the past three decades, they all shared with us their knowledge, professional experience, thoughts and attitudes with respect to computer science education.

September 2014

Orit Hazzan
Tami Lapidot
Noa Ragonis

Index of Activities

The activity page numbers appear in the table of Contents

Chapter 3 Overview of the Discipline of Computer Science

		Section
1 The Nature of Computer Science	A. Explain what computer science is, work in pairs	3.2
	B. Class discussion	
	C. Internet exploration of computer science definitions	
	D. Summary and class discussion	
	E. Review of the Computing Curricula 2001, homework	
2 Computer Science and Other Sciences	A. Connections between computer science and other sciences, individual/ team work	3.2
	B. Presentations	
	C. Class discussion	
3 Plan a Lesson about the History of Computer Science	A. Introductory questions	3.3
	B. Plan a lesson, work in pairs	
	C. Presentations	
	D. Class discussion	
4 History of Computational Machines		3.3
5 Preparation of a Presentation on a Computer Scientist, Homework		3.4
6 Analysis of Ethical Dilemmas	A. Case analysis, group work	3.5.1
	B. Presentations and discussion	
7 Diverse Class Demography, Group Work		3.5.2
8 Test Evaluation, Work in Pairs		3.5.2
9 Gender Diversity, Open Conversation		3.5.2
10 Introduction to Soft Skills, Homework towards the Lesson on Soft Skills		3.5.3

Chapter 4 Research in Computer Science Education

		Section
25 Exploration of Computer Science Education Research	Work on learners' understanding of basic computer science topics	4.3
26 The Computer Science Education Research World	A. Intuitive thinking on computer science education research, class discussion	4.3
	B. Planning research in computer science education, group work	
	C. Class discussion	
27 Looking into Research Work on Novices' Difficulties, Homework		4.3
28 The Teacher as a Researcher	A. Solving a problem, individual work	4.3
	B. Evaluating different solutions, individual work	
	C. Discussion on Stage B answers, work in pairs	
	D. The meaning of learners' mistakes, class discussion	
	E. Taking the researcher's perspective, work in pairs	
	F. Reflection, individual work	
29 Reflection on Reading a Computer Science Education Paper, Homework		4.3

Chapter 5 Problem-Solving Strategies

		Section
30 Problem-Solving Techniques in Computer Science		5.2
31 Examination of Representative Inputs and Outputs	A. Problem development, work in pairs	5.3
	B. Presentations and discussion	
32 Choosing the Problem Variables	A. Problem analysis, work in pairs	5.4.1
	B. Discussion between pairs	
	C. Presentations and discussion in the course plenum	
33 Roles of Variables—Discovery Learning and Reflection	A. Learning, work in pairs	5.4.1
	B. Reflection on Stage A, individual work	
34 Roles of Variables—Examination of the Roles of Variables Through the Research Lens		5.4.1
35 Practicing Stepwise Refinement— Break Down Problem Solutions into Subtasks		5.4.2

Chapter 6 Learners' Alternative Conceptions

Chapter 7 Teaching Methods in Computer Science Education

		Section
47 Pedagogical Examination of Games	A. Playing a game	7.2.1
	B. Class discussion	
	C. Game design	
	D. Presentation and class discussion	
	E. Design and construct a game	
	F. Playing the games	
	G. Summary	
48 Educational Usage of Games in Computer Science Education		7.2.1
49 Pedagogical Examination of the CS-Unplugged Approach	A. Experience a CS-Unplugged activity	7.2.2
	B. Exploration of the CS-Unplugged approach	
	C. Design of a CS-Unplugged activity, work in pairs	
	D. The CS-Unplugged approach and other computer science teaching methods, homework	
50 Pedagogical Examination of Rich Tasks	A. Solving a rich task, individual work	7.2.3
	B. Presentation of learners' solutions	
	C. Class discussion	
	D. Construction of a rich task, homework	
51 Pedagogical Examination of Concept Maps	A. Concept map construction, group work	7.2.4
	B. Concept map evaluation, group work	
	C. Class discussion	
52 Pedagogical Examination of Classification	A. Classification activity, group work	7.2.5
	B. Class discussion	
	C. Construction of a classification activity, homework	
53 Metaphors—Preparing a Poster, Variable Exhibition		7.2.6
54 Metaphors—Advantages and Disadvantages of Metaphors		7.2.6
55 Different Forms of Class Organizations		7.3
56 Distance Learning in the School—The Flipped Classroom	A. Homework towards the lesson on the flipped classroom	7.3
	B. Teamwork in class	

Chapter 8 Lab-Based Teaching

Chapter 9 Types of Questions in Computer Science Education

Chapter 10 Assessment

Chapter 11 Teaching Planning

Chapter 12 Integrated View at the MTCS Course Organization: The Case of Recursion

Contents

Introduction—What Is this Guide About?

1

Abstract

This chapter presents the motivation for writing this guide, the Methods of Teaching Computer Science (MTCS) course, for which the guide can serve as a textbook, the structure of the guide, and how it can be used in different frameworks of computer science education.

1.1 Introduction

This guide is about computer science teaching. Specifically, it focuses on the Methods of Teaching Computer Science (MTCS) course, in which high school computer science teachers get their pedagogical education related to computer science teaching. Though the guide is organized as a textbook for the MTCS course, most of its ideas can be easily adapted to the teaching of any computer science topic in any framework and any level, from middle school through high school to the university level. According to Abramson (2011), "This book is directed at those who teach computer science (CS) in high schools or undergraduate classrooms.…[and] is a valuable resource for high school and undergraduate teachers of introductory CS courses."

Accordingly, the computer science teaching aspects presented in this guide are not restricted to any specific curriculum and can be applied in the teaching of any subject, including Advanced Placement (AP) contents. However, in order to provide a common base for the entire guide readership, most of the illustrations presented in this guide are based on fundamental computer science concepts; nevertheless, they can be easily adjusted to any other computer science topic. Section 1.5 specifies how this guide can be used by different populations of computer science educators: instructors of the MTCS course, computer science instructors in the university, instructors of in-service teacher professional development programs, and high school computer science teachers.

© Springer-Verlag London Limited 2014
O. Hazzan et al., *Guide to Teaching Computer Science*,
DOI 10.1007/978-1-4471-6630-6_1

This variety of readership is also addressed by using the following terminology throughout this guide:

- *Learners*: Computer science learners in any framework, either in the university or in the middle and high school.
- *Students*: Prospective high school computer science teachers, that is, the students enrolled in the MTCS course. When it is important to indicate that the students are learning toward their high school teaching certificate in computer science, we call them *prospective high school computer science teachers*.
- *Pupils*: High school computer science learners.

In this chapter, we also present the motivation for writing the guide (Sect. 1.2); the MTCS course, for which the guide can serve as a textbook (Sect. 1.3); the structure of the guide (Sect. 1.4); and how it can be used in different frameworks of computer science education (Sect. 1.5).

1.2 Motivation for Writing this Guide[1]

The dynamic evolution of the computer science field poses also educational and pedagogical challenges, including computer science teacher-related issues, such as recruitment, pre-service teacher preparation, support for teachers' ongoing professional development, and pedagogical and instructional design of teaching and learning material (Stephenson 2005).

In this context, The ACM K-12 Education Task Force Report draws attention to the need for appropriate computer science teacher training programs and notes that "teachers must acquire both a mastery of the subject matter and the pedagogical skills that will allow them to present the material to students at appropriate levels" (Tucker et al. 2007, p. 18). However, according to a recently published report, "Despite the existence of National Council for the Accreditation of Teacher Education accreditation requirements for computer science, very few pre-service teacher preparation programs have the current capacity or coursework developed to prepare computer science teachers" (Wilson et al. 2010, p. 12).

Nevertheless, in many places, a computer science teaching certificate is not required in order to teach computer science. In the USA, for example, a survey conducted in 2007 reports that approximately 53 % of the respondents replied that their state does not require a computer science teaching certification (CSTA 2007). Further, even programs that deal specifically with the training of computer science teachers do not necessarily include explicit reference to the teaching of computer science. Rather, in many cases, the training refers to teaching in general and, at

---

[1] Based on Ragonis and Hazzan (2008) Disciplinary-pedagogical teacher preparation for pre-service Computer Science teachers: Rationale and implementation, Informatics in Secondary Schools—Evolution and Perspective—ISSEP 2008, Lecture Notes in Computer Science, Vol. 5090/2008: 253–264. Included with permission here.

the best, to principles of science teaching. This might result from the fact that a well-defined international standard for computer science school curricula and for computer science teacher preparation does not exist (Ragonis 2009).

To meet the challenge of preparing future computer science teachers, this guide can be used either as a teaching guide or as a textbook (or both) for the MTCS course. Specifically, the guide presents a conceptual framework, together with detailed implementation guidelines, for general computer science teaching situations as well as for the MTCS course. Practically, its writing style enables immediate implementation of its ideas in the MTCS course as well as in other computer science teaching education frameworks.

From a personal perspective, our own motivation for writing this guide is based on three decades of teaching experience, management, and involvement in teacher preparation programs. Anecdotally, we mention that when we started building the MTCS course in our relative institutions around 30 years ago, it was almost impossible to find any model for high school computer science teacher preparation program around the world. Thus, in fact, this guide enables us to share with the professional community of computer science educators the accumulated professional knowledge we have gained over these years.

1.3   MTCS Course

This section presents the Methods of Teaching Computer Science course—the MTCS course as it is referred to in this guide. We present a general overview of the knowledge structure required from pre-service computer science teachers, the course population, the course rationale, including detailed objectives, and recommended teaching methods to be used in the course. Chapter 14 further describes how to design an MTCS course and suggests two possible syllabi for the course.

1.3.1   MTCS Course Overview

Teacher preparation programs are usually based on general pedagogical knowledge, subject matter knowledge, pedagogical content knowledge (PCK), and practicum in real classes. PCK is one category of Shulman's Teacher Knowledge Base Model (Shulman 1986), and it refers to what a teacher is required to know in order to teach a certain subject matter: how to make it understandable, learners' preconceptions and misconceptions, and strategies for coping with learners' misconceptions (Shulman 1986, 1990).

From this perspective, the MTCS course aims at broadening the students' PCK and sets the basis for the in-school training that takes place after it. In other words, based on the working assumption that the computer science prospective teachers learn the discipline of computer science and general pedagogy topics in other courses, the MTCS course focuses on the uniqueness of teaching computer science. In this context, Gal-Ezer and Harel (1998) claim that "beyond the mastery of core computer science material, good computer science educators should also be familiar with a significant body of material that will expand their perspectives on the field, and consequently, enhance the quality of their teaching" (p. 77). Among the issues they mention are the question what is computer science?, a bird's-eye view of the discipline, and familiarity with teaching tools and methods.

Lapidot and Hazzan, in a series of papers, address these issues practically and discuss different topics related to computer science teacher preparation in general and to the MTCS course in particular (Hazzan and Lapidot 2004a, b, 2006; Lapidot and Hazzan 2003, 2005). They refer to different topics that should be included in such a course, like pedagogical approaches for teaching different subjects, tools for assessing pupils' performance, and teaching of social issues such as ethics. They also emphasize the need to use active learning when teaching the MTCS course. Hazzanet et al. (2008) add that computer science teacher preparation programs should include some research elements, such as reading assignments of papers that deal with computer science education research and mini-research projects to be carried out by the prospective computer science teachers themselves.

In recent years, several new publications about the MTCS course were published, as is described in what follows.

According to Armoni (2011), the methods of course should serve as a bridge between subject matter knowledge (SMK), pedagogical knowledge (PK), PCK, and curricular knowledge. Armoni recommends that the MTCS course should include specific PCK, web-based scientific inquiry, various ways of asking questions, prospective teachers' beliefs, problem-based learning. In general, it should aim at integrating all kinds of knowledge, as well as beliefs, and should incorporate concreteness, whether by involving a field-experience component or by using other tools, such as case studies and problem-based learning. The course should provide opportunities for collaboration, and reflection should be extensively employed throughout the course.

According to Yadav and Korb (2012), a methods course is typically where prospective teachers learn about "pedagogical ways of doing, acting, and being a teacher." They argue that high school CS teachers must have in-depth computer science knowledge as well as strong PCK, developed through a MTCS course. Therefore, a MTCS course should be "about how CS is learned and taught, and about how classrooms can provide an environment for learning CS."

They introduce a MTCS course, taught at Purdue University, that develops students' PCK through experiences that allow them to think and act like computer science teachers: understand ways of representing and formulating the subject matter and make it understandable to students, know which topics students find easy or difficult to learn, which ideas (often misconceptions) students bring with them to the classroom, and how to transform those misconceptions, understand how students develop and learn, and how to teach diverse learners. Specifically, their course trains prospective computer science teachers to combine pedagogical principles with computer science content to improve the learning experience for their students. The course involves reading, discussing, and reflecting on papers that describe pedagogical practices of teaching computer science principles.

Zimudzi (2012) also suggests an MTCS course in which there are four general objectives:

- Demonstrate knowledge of programming techniques
- Apply algorithmic tools in solving problems
- Show understanding of different programming languages
- Show understanding of program translators

Gal-Ezer and Zur (2013) present a computer science teacher certification program being offered by the Israeli Open University, combined of courses and practical training. They describe their version of the methods course, which is called *Topics in Computer Science Education*. This course was designed to combine several academic skills: self-study of professional scientific literature, scientific research skills, presentation skills, and the skills required for preparing and delivering a lesson.

Another suggestion for the MTCS course was piloted in the Center X Teacher Education program at the University of California, Los Angeles (UCLA) as part of the MOBILIZE project[2]. The two-unit course explores CS "as a discipline that encourages inquiry, creativity, and collaboration" and highlights "four related areas of emphasis: (1) Representing and connecting CS concepts; (2) Engagement with particular instructional strategies that foster inquiry-based teaching and learning; (3) Eliciting, assessing, and understanding students' CS content knowledge and computational thinking practices; (4) Development of equity-based teaching practices in CS education."

This guide elaborates on all the above mentioned topics as well as on a variety of additional topics.

1.3.2   Course Population

Course participants are either prospective or in-service computer science teachers who wish to broaden their education related to computer science education in the high school. In addition, some students study the course even though they do not intend to become computer science teachers. These students can apply the pedagogical skills they acquire in the course in their future professional work either in the academia or the industry. This wide perspective at computer science teaching includes also those who view computer science teaching as an additional profession, which they may use at some point in the future (Hazzan and Ragonis 2014).

The course's academic prerequisites are the relevant computer science courses and part of the general education and teaching studies, e.g., psychology.

---

[2] Computer Science Project: http://centerx.gseis.ucla.edu/parent-project.

1.3.3   Course Objectives

The MTCS course is one of the first stages in the professional development of the prospective computer science teachers, which naturally continues in many ways also after the course ends (e.g., in their participation in conferences, in-service teachers training programs, working groups of computer science teachers, design of new curricula, and more). Accordingly, the main objective of the MTCS course is to prepare the students to their future fieldwork as computer science teachers. The specific objectives derived from this main objective, are:

1. To enhance students' professional identity as computer science teachers
2. To heighten students' awareness to the uniqueness of computer science education
3. To expose students to difficulties encountered by learners when learning different topics from the computer science curriculum
4. To enable students to master pedagogical skills for teaching computer science, considering different kinds of learners
5. To enable students to master pedagogical tools for teaching computer science, including the creation of a supportive and cooperative inquiry-based learning environment
6. To expose the students to a variety of computer science teaching methods
7. To expose students to the research conducted on computer science education and to its application in the teaching process

These objectives are achieved by the facilitation of many activities, discussions, reflective thinking processes, and additional kinds of tasks, as is elaborated in the next section and is widely illustrated in this guide.

1.3.4   Recommended Teaching Methods Used in the MTCS Course

The teaching methods implemented in the MTCS course are varied and their implementation in itself constitutes an essential tier of learning the course. The course does not only "talk about" but rather it "shows how" to actively apply the teaching principles in the teaching of the discipline of computer science. Thus, the MTCS course is essentially a workshop that includes lectures, workshops for developing different teaching materials and skills, hands-on experience with various software programs, practice teaching, and many discussions and reflections. Course tasks and assignments are varied and develop simultaneously with the learning process. Task types include, for example, preparing learning activities, preparing lesson plans, analyzing learners' mistakes, reading articles and preparing reports, exploring different ways for class management in general and how to use the computer lab in particular, preparing a teaching plan for an entire teaching unit, and exploring the uniqueness of the discipline of computer science. These kinds of activities, as well as many others, are presented in the guide with respect to different (pedagogical) computer science topics.

Furthermore, the teaching-learning processes in the course are cooperative in nature. This idea is applied by letting students present different products to their peers and learn from one another and from feedback they receive from their peers and course instructor. Feedback can be given both orally and in writing. The products of the course participants can be shared for future use either in the course or in the schoolwork.

It is recommended to accompany the course with a website that includes available resources throughout the school year. These resources include links to repositories of learning materials and to sites that offer enrichment on topics such as the history of computer science, dictionaries of programming languages, information on computer scientists, and so on. In addition, all contents presented during the course lessons both by the instructor and by the students can become available on the Web.

Chapter 2 further elaborates on the teaching methods employed in the course.

1.4   The Structure of the Guide to Teaching Computer Science

This section describes the structure of the guide and briefly presents each of its chapters.

1.4.1   Guide Structure and Organization

As mentioned, this guide can be used as a textbook for the MTCS course as well as a guide for computer science teaching in a variety of frameworks, organizations, and levels of computer science education. To achieve this purpose, each chapter presents:

- A pedagogical computer science aspect and includes its meaning and importance in the context of computer science education, its basic pedagogical principles, and how it can be applied in computer science education in general and in the MTCS course in particular.
- Examples of activities that address the pedagogical aspect the chapter deals with. The learning activities are presented in detail, in gray boxes, including pedagogical guidelines, to enable their immediate application. In order to make the material applicable to all levels of computer science education, we do not indicate explicitly the length of each activity; each teacher/instructor, we suggest, allocate the time framework for each activity according to the characteristics of his or her learners. We also include an Index of Activities at the beginning of the guide.

We highlight that:

- The guide does *not* teach computer science; rather, it focuses on computer science *teaching*.

- The guide is *not* limited to the teaching of a specific computer science curriculum, neither is it limited to the teaching of a specific programming language nor to a specific programming paradigm.
- The pedagogical topics that the guide deals with should not necessarily be addressed by their presentation order in the guide.
- The programming language we use in the guide is Java, since currently it is one of the most common languages. The Java code can be translated, of course, to any other programming language.
- The guide contains much more material than a single MTCS course can contain. Therefore, each instructor should select the topics and activities that fit the context in which the MTCS course is taught or, in other computer science teaching frameworks, the specific purpose for which the guide is used.
- When the guide is used as the textbook of the MTCS course, it should be remembered that the course should be based on active learning (see Chap. 2); that is, computer science teaching *cannot* be learnt by reading this guide.
- Chapter 14 presents two optional syllabi for the MTCS course. However, many options exist for the teaching of the MTCS course and, accordingly, the material presented in this guide can be organized and implemented in additional ways (see, e.g., Chap. 12, which presents a course organization around one central idea—recursion).
- Each lesson of the MTCS course can be organized in many ways. Activity 110 (Sect. 14.2.2) illustrates several options as to how to start the first lesson of the MTCS course.

1.4.2   The Content of the Guide Chapters

In what follows, we briefly describe the content of each chapter of the guide.

Chapter 2—Active Learning and the Active-Learning-Based Teaching Model. This chapter presents an active learning-based teaching model for implementation in the MTCS course. This model is applied in the various chapters of this guide in most of the offered activities.

Chapter 3—Overview of the Discipline of Computer Science. This chapter addresses topics associated with the nature of the discipline of computer science and with cross-curriculum topics. The importance of these topics is explained by the fact that even today no consensus has been reached with respect to one agreed upon definition for computer science, and different scholars view it differently. Specifically, the following topics are included in this chapter: what is computer science?, the history of computer science, computer scientists, social issues of computer science, programming paradigms, computer science soft ideas and skills, interdisciplinary aspects of computer science and big data.

Chapter 4—Research in Computer Science Education. This chapter focuses on research in computer science education. The importance of including this topic in the MTCS course stems from the fact that computer science education research can enrich the prospective computer science teachers' holistic perspective with respect to the discipline of computer science, the computer science teacher's role, ideas and ways to apply different teaching methods, and students' difficulties, misconceptions, and cognitive abilities. In practice, this knowledge may foster their professional development and enhance their future teaching with several respects, such as, lesson preparation, activities developed for their pupils, teacher's behavior in the class, and testing and grading learners' projects and tests.

Chapter 5—Problem-Solving Strategies. The importance of including this topic in the MTCS course stems from the centrality of problem-solving processes in computer science and the fact that computer science learners often experience difficulties in the problem analysis and solution construction processes. This chapter deals with pedagogical tools needed to be acquired by computer science teachers to help their pupils acquire different problem-solving skills, such as successive refinement, the use of algorithmic patterns, testing and debugging, and reflective processes.

Chapter 6—Learners' Alternative Conceptions. Since alternative conceptions are not easily detected by conventional testing and evaluation methods, a teacher cannot effectively deal with alternative conceptions without being aware of their existence. More specifically, teachers must be aware of their learners' ways of thinking and mental processes, must gain skills for uncovering alternative conceptions, and must recognize and use pedagogical tools to deal with these conceptions. In order to prepare the prospective computer science teachers master these skills, one of the messages delivered in this chapter is that a learning opportunity exists in every pupil's mistake (or misunderstanding); in order to exhaust pupil's learning abilities, however, it is necessary first, to understand the pupil's (alternative) conceptions and then, to use suitable pedagogical tools to assist him or her.

Chapter 7—Teaching Methods in Computer Science Education. This chapter includes active-learning-based teaching methods that computer science teachers can employ in their classroom. The purpose of this chapter is first, to let the students in the MTCS course experience a variety of teaching methods before becoming computer science teachers; second, to discuss, together with the students, the advantages and disadvantages of these teaching methods; and third, to demonstrate high school teaching situations in which it is appropriate to employ these teaching methods. The teaching methods presented in this chapter are pedagogical tools, different forms of class organization, mentoring software project development, massive online open courses (MOOCs), and mobile learning.

Chapter 8—Lab-Based Teaching. This chapter focuses on computer science teaching methods that fit especially to be employed in the computer lab. The importance of the computer lab as a learning environment for computer science is explained by the fact that the lab enables learners to practice and explore problem-solving strategies, to express their solutions to a given problem and to get the computer immediate feedback, and to deepen their understanding of algorithms they develop. One of the aims of this chapter is to let the prospective computer science teachers realize that the learning of computer science in the computer lab is not limited to programming tasks; rather, they, as future computer science teachers, can use the computer lab in additional ways that further enhance learners' understanding of computer science. Specifically, this chapter includes the following topics: what is a computer lab?, the lab-first teaching approach, visualization and animation, and using the Internet in the teaching of computer science.

Chapter 9—Types of Questions in Computer Science Education. This chapter explores and discusses the variety of question types that a computer science teacher can use in different teaching situations and processes: in the classroom, in the computer lab, as homework, and in tests. The integration of different types of questions has several pedagogical advantages. We mention four: first, different types of questions enable to illuminate different aspects of the learned content; second, different types of questions require the students to use different cognitive skills; third, different types of questions enable the teacher to vary his or her teaching tools; and fourth, the integration of different types of questions throughout the teaching process keeps the students' interest, attention, and curiosity. It is important to address this theme is the MTCS course to increase the prospective computer science teachers' awareness to the fact that the use of different types of problems in their teaching processes can enrich their pupils' variety of thinking processes and expand the spectrum of their cognitive skills. Special attention is given to problem-solving questions, their formatting (for example, keywords used), and the needed skills to solve them.

Chapter 10—Assessment. Assessment is one of the most common tasks teachers perform from the early stages of their professional development. This chapter highlights the uniqueness of assessing learners' products and learning processes in the case of computer science education. The main message conveyed in this chapter is that assessment is not a target by itself, but rather, a pedagogical means by which, learners get feedback related to their own understanding of the learned topic, and the teacher learns on the understanding of the current knowledge of his or her pupils, and can accordingly improve his or her teaching tools and methods to help them overcome their current obstacles. The topics presented in this chapter are test construction and evaluation, project evaluation, and the use of a portfolio in computer science education.

Chapter 11—Teaching Planning. Planning a teaching sequence is one of the basic stages of any teaching process. This chapter illustrates a top-down approach for teaching planning. It starts with a broad perspective that relates to the planning of an entire curriculum (e.g., an introductory computer science course), continues with the planning of one topic from the curriculum (e.g., teaching one-dimension array), and finally addresses the planning of a specific lesson (e.g., the first lesson about arrays). In all these stages, the multifaceted considerations that a teacher should be aware of while planning the teaching process are addressed. In addition, an approach to teaching planning is presented, according to which a complicated concept understanding is constructed in a spiral gradient manner.

Chapter 12—Integrated View at the MTCS Course Organization: The Case of Recursion. This chapter reviews the guide's chapters systematically through the lens of recursion—one of the central computer science concepts. The main message of the chapter is to demonstrate that the entire course can be organized around one computer science core idea. The ideas presented in this chapter can be used by instructors of the MTCS course as well as by other computer science educators. We mention, though, that recursion is only one candidate for such course organization and other computer science concepts, such as abstraction, control structures, and abstract data types, can be used for this purpose.

Chapter 13—Getting Experience in Computer Science Education. This chapter deals with the first field teaching experiences that the students enrolled in the MTCS course gain before becoming computer science teachers. The importance of these first teaching experiences stems from the recognition that one significant way to acquire pedagogical-disciplinary knowledge involves activities performed in teaching situations that provide opportunities for teacher's reflective processes. This chapter describes three frameworks in which the prospective computer science teachers gain this first teaching experience: The practicum, which takes place in high school, after one or two semesters of learning the MTCS course, CS teacher training within the professional development school (PDS) collaboration framework, and a tutoring framework that can be integrated in the MTCS course.

Chapter 14—Design of a Methods of Teaching Computer Science course. This chapter describes how to design a MTCS course within an academic computer science teacher preparation program, and suggests two possible syllabi for this course. It is emphasized, however, that different approaches and frameworks can be applied when one designs the MTCS course. In the first part of this chapter, we propose four possible perspectives on the MTCS course: the NCATE standards, merger of computer science with pedagogy, Shulman's model of teachers' knowledge, and research findings. The second part of the chapter describes two full course syllabi for the MTCS course. We also present the approach which offers computer science students an additional profession—computer science teachers.

Chapter 15—High School Computer Science Teacher Preparation Programs. This chapter broadens the perspective on high school computer science teacher preparation programs. First, it describes a model for high school computer science education that one of its components is computer science teacher preparation programs. The model consists of five key elements—a well-defined curriculum, a requirement of a mandatory formal computer science teaching license, teacher preparation programs, national center for computer science teachers, and research in computer science education—as well as interconnections between these elements. Then, the focus is placed on the teacher preparation programs component of the model, describing a workshop targeted at computer scientists and computer science curriculum developers who wish to launch computer science teacher preparation programs at their universities but lack knowledge about the actual construction of such programs.

1.5   How to Use the Guide?

As mentioned, this guide can be used by several populations involved in computer science education. We now present these different usages.

1.5.1   Instructors of the MTCS Course

When this guide is used as a textbook for the MTCS course, it serves, in fact, as a teaching guide. In this case, the instructor of the MTCS course can use a two-

Table 1.1 Organization of the contents of the MTCS course

Computer science contents	Variables	Control structures	Recursion	...	And so forth—additional computer science topics
Pedagogical aspects					
Learners' difficulties and conceptions		+	+		
Teaching methods in computer science education		+			
Lab-based teaching	+				
Evaluation	+				
...					
And so forth—other pedagogical aspects					

dimensional table, such as Table 1.1, in order to organize the course structure and contents. The columns of the table represent the main contents and ideas of the high school curriculum on which the MTCS course focuses; the table rows present the pedagogical contents and ideas that the instructor wishes to address in the course. Thus, the different chapters of this Guide represent, in fact, the table rows.

During the actual teaching of the MTCS course, the instructor decides what pedagogical ideas to address with respect to what computer science topic(s). He or she can choose, for example, to discuss the *computer science topic* of variables with respect to the *pedagogical aspects* of evaluation and lab-based teaching; the *computer science topic* of control structures—with respect to the *pedagogical topics* of learners' difficulties and conceptions and teaching methods in computer science education, etc. In this way, each activity in the course is associated with both a computer science topic and a pedagogical aspect. It also implies that none of the lessons of the MTCS course focuses solely on a computer science idea (without a pedagogical context) or on a pedagogical idea (without connecting it to the teaching of some computer science topic). In this spirit, though the pedagogical aspects presented in this guide are presented in most cases in the context of a specific computer science topic, they can be easily adapted for the teaching of other computer science topics. Section 10.5 addresses the evaluation of the students in the MTCS course and outlines possible elements for students' evaluation in the course.

By the end of the course, such a table should not be full; it reflects, however, whether all the computer science topics included in the curriculum have been discussed in the MTCS course and if all the pedagogical ideas the instructor wished to highlight were indeed addressed. Nevertheless, it is recommended to address all computer science topics and all the pedagogical aspects; the specific topic selection and their presentation order in the course is left to the course instructor according to his or her pedagogical preferences.

1.5.2   The Prospective Computer Science Teachers Enrolled in the MTCS Course

The students enrolled in the MTCS course can use the guide as they use any other textbook. The instructor can refer them to the general introduction presented for each topic (presented in most cases at the beginning of any chapter/section) or to specific assignments, according to the course plan. Interested students can, of course, read additional material and further broaden their knowledge by reading the references presented at the end of each chapter.

1.5.3   Computer Science Instructors in the University

Computer science instructors can use the guide in several ways.

First, they can expand their knowledge related to computer science education and improve their awareness to pedagogical aspects while teaching computer science courses.

Second, the guide may increase their awareness to alternative conceptions and possible difficulties that learners in their classes may face with respect to different computer science topics.

Third, they can facilitate with their classes activities presented in the guide, even if their students do not intend to become computer science teachers. Naturally, in such cases, the computer science context should be emphasized.

Finally, computer science instructors can use this guide to vary their teaching methods in order to promote learners' interest and motivation. This can be done, for example, by integrating active-learning-based activities in the lessons they teach.

1.5.4   Instructors of In-Service Teachers' Professional Development Programs

Most of the topics and activities presented in the guide are also suitable for facilitation with in-service computer science teachers in different frameworks, such as professional development training and workshops in conferences.

1.5.5   High School Computer Science Teachers

Similar to instructors of computer science courses in the university, the guide can be used by high school computer science teachers in several ways. They can expand their knowledge related to computer science education, improve their computer science teaching, vary the teaching methods they employ in their classes, and increase their awareness to possible difficulties their pupils may face with respect to different computer science topics. They can also facilitate with their high school pupils activities presented in this guide.

References

Abramson G (September, 2011) ACM computing reviews. http://www.amazon.com/Guide-Teaching-Computer-Science-Activity-Based/dp/0857294423. Accessed August 2014

Armoni M (2011) Looking at secondary teacher preparation through the lens of computer science. ACM Trans Comput Educ 11(4), Article 23, 23:1–23:38

CSTA (2007) Compute Science State Certification Requirements—CSTA certification committee report http://www.csta.acm.org/ComputerScienceTeacherCertification/sub/TeachCertRept-07New.pdf. Accessed 14 July 2010

Gal-Ezer J, Harel D (1998) What (else) should CS educators know? Communic ACM 41(9):77–84

Gal-Ezer J, Zur E (2013) What (else) should CS educators know?—revisited. *WiPSCE '13*, November 11–13 2013, Aarhus, Denmark. pp. 84–87

Hazzan O, Lapidot T (2004a) Construction of a professional perception in the "Methods of Teaching Computer Science" course. Inroads—SIGCSE Bull 36(2):57–61

Hazzan O, Lapidot T (2004b) The practicum in computer science education: bridging gaps between theoretical knowledge and actual performance. Inroads—SIGCSE Bull 36(4):47–51

Hazzan O, Lapidot T (2006) Social issues of computer science in the "Methods of Teaching Computer Science in the High School" course. Inroads—SIGCSE Bull 38(2):72–75

Hazzan O, Ragonis N (2014) STEM teaching as an additional profession for scientists and engineers: the case of computer science education, Proceedings of SIGCSE 2014—The 45th ACM technical symposium on computer science education, Atlanta, GA, USA: 181–186

Hazzan O, Gal-Ezer J, Blum L (2008) A model for high school computer science education: the four key elements that make it!, 39th Tech. Symp. on Comput. Sci. Educ. SIGCSE Bull 40(1):281–285

Lapidot T, Hazzan O (2003) Methods of teaching computer science course for prospective teachers. Inroads—SIGCSE Bull 35(4):29–34

Lapidot T, Hazzan O (2005) Song debugging: merging content and pedagogy in computer science education. Inroads—SIGCSE Bull 37(4):79–83

Ragonis N (2009) Computing pre-university: secondary level computing curricula. In Ed. Benjamin W. Wah. (eds): Wiley Encycl. of Comput. Sci. and Eng: Ed. Benjamin W. Wah. 5(1), pp. 632-648. John Wiley & Sons, Inc., Hoboken, NJ, USA.

Ragonis N, Hazzan O (2008) Disciplinary-pedagogical teacher preparation for pre-service computer science teachers: rational and implementation, ISSEP 2008, lect. notes in comput. Sci. 5090/2008: 253–264

Shulman LS (1986) Those who understand: knowledge growth in teaching. Educ Teach 15(2):4–14

Shulman LS (1990) Reconnecting foundations to the substance of teacher education. Teach Coll Rec 91(3):300–310

Stephenson C, Gal-Ezer J, Haberman B, Verno A (2005) The new educational imperative: improving high school computer science education. Final report of the CSTA curriculum improvement task force February 2005, Comput. Sci. Teach. Assoc., Assoc. for Comput. Mach. http://www.csta.acm.org/Communications/sub/DocsPresentationFiles/White_Paper07_06.pdf. Accessed 14 July 2010

Tucker A, Deek F, Jones J, McCowan D, Stephenson C, Verno A (2007) A model curriculum for K-12 computer science. Report of the ACM K-12 Educ. Task Force Comput. Sci. Curric. Comm.—Draft http://www.csta.acm.org/Curriculum/sub/CurrFiles/K-12ModelCurr2ndEd.pdf. Accessed 14 July 2010

Wilson C, Sudol L, Stephenson C, Stehlik M (2010) Running on empty: the failure to teach K–12 computer science in the digital age. Report of the Assoc. for Comput. Mach. And The Comput. Sci. Teach. Assoc. http://www.acm.org/runningonempty/fullreport.pdf. Accessed 8 Oct 2010

Yadav A, Korb JT (2012) Learning to teach computer science: the need for a methods course: a multipronged approach to preparing computer science teachers is critical to success. Commun ACM 55(11):31–33

Zimudzi E (2012) Active learning for problem solving in programming in a computer studies method course. Acad Res Int 3(2):284–292

Active Learning and Active-Learning-Based Teaching Model

Abstract

This chapter presents an active-learning-based teaching model for implementation in the Methods of Teaching Computer Science (MTCS) course, which is based on the constructivist approach. This model is used in most of the offered activities in this guide. The chapter starts with motivation and rationale for using active learning in the MTCS course; then, the active-learning-based teaching model is introduced and explained, including a description of the role of the instructor of the MTCS course in the model implementation.

2.1 Introduction

As mentioned in the Introduction (Chap. 1), the main purpose of the Methods of Teaching Computer Science (MTCS) course is to prepare prospective computer science teachers (the students of the course) toward their future career as computer science teachers.

In general, courses about science teaching in the secondary school emphasize curriculum-related issues, addressing topics such as learning theories and pedagogical methods, principles for the development of scientific curricula, laboratory instruction and other investigative learning approaches, professional ethics in science instruction, and the place of science learning in the pupils' general education. All these topics are also relevant and important in the case of computer science teaching to promote the prospective teachers' professional perception.

The recommended teaching methods for the MTCS course, as described in Chap. 1, indicate that the MTCS course should be built as a teaching model. Accordingly, the course should be designed in a way that (a) promotes students' positive learning experience in a supportive teaching environment and (b) enables the

---

© Springer-Verlag London Limited 2014
O. Hazzan et al., *Guide to Teaching Computer Science*,
DOI 10.1007/978-1-4471-6630-6_2

students to imitate this way in their future computer science classes. To achieve this goal, the MTCS course should be based on constructivist teaching methods and implement active learning. This approach is important not only because we want the prospective computer science teachers to enjoy their learning processes and improve their understanding of computer science concepts, science teaching, and computer science education (by experiencing a variety of learning/teaching methods), but also because we want to inspire their future way of teaching in the high school.

This chapter presents an active-learning-based teaching model for implementation in the MTCS course. This model is used in this guide in most of the offered activities.

2.2 Active Learning

Confucius (551 BC—479 BC) once said:

I hear and I forget,
I see and I remember,
I do and I understand.

Active learning is widely accepted nowadays as a quality form of education. Among the many descriptions of active learning, we highlight Silberman's assertion (1996) according to it "Above all, students need to 'do it'—figure things out by themselves, come up with examples, try out skills, and do assignments that depend on the knowledge they already have or must acquire." Active learning is closely related to inquiry-based learning, problem-based learning, and project-based learning, which are all highly suitable approaches[1] for computer science learning in general and for the MTCS course in particular.

According to constructivist educators (Kilpatrick 1987; Davis et al. 1990; Confrey 1995), learning is an active acquisition of ideas and knowledge construction, rather than a passive process. In other words, learning requires the individual to be active and to be engaged in the construction of one's own mental models. As follows from the above quote by the famous Chinese philosopher, the more active learners are, the more meaningful is their understanding of what they learn. Therefore, in the design of the MTCS course, we propose educators to encourage "learners to be active in their relationship with the material to be learned" (Newman et al. 2003).

There are numerous ways to implement active learning in computer science education (see, e.g., Whittington 2004; Ludi 2005; McConnell 2005; Anderson et al. 2007; Gehringer and Miller 2009; Walker 2011; Zimundi 2012). McConnell (1996), for example, suggests several techniques, such as modified lectures, algorithm tracing, and software demonstration. In this spirit, this guide is based on the implementation of the active-learning-based teaching approach by offering a wide collection of activities to be implemented in the MTCS course in the context of computer science education.

---

[1] A good resource on active learning, including different types of activities, can be found in http://www1.umn.edu/ohr/teachlearn/tutorials/active/what/index.html**

2.3   Why Active Learning Is Suitable for Implementation in the MTCS Course?

In addition to the general argumentation about the suitability of the active-learning-based teaching approach to the MTCS course, we suggest that active learning may also promote the professional development and perception of the prospective computer science teachers, as the following justifications propose.

- *Constructivism*: Constructivism is a cognitive theory that examines the nature of learning processes. According to this approach, learners construct new knowledge by rearranging and refining their existing knowledge (cf. Davis et al. 1990; Smith et al. 1993; Ben Ari 2001). More specifically, the constructivism approach suggests that new knowledge is constructed *gradually*, based on the learner's existing mental structures and on the feedback that the learner receives from the learning environments. In this process, mental structures are developed in steps, each elaborating on the preceding ones, although there may, of course, also be regressions and blind alleys. This process is closely related to the Piagetian mechanisms of assimilation and accommodation (Piaget 1977). One way to support such gradual mental constructions is by providing learners with a suitable learning environment in which they can be *active*. The working assumption is that the feedback, provided by learning environment in which learners learn a complex concept in an active way, may support mental constructions of the learned concepts. In our case, in order to support the construction of the computer science teachers' professional perception, the prospective teachers participating in the MTCS course must have a learning environment that supports this complex mental construction. It is suggested, therefore, that active learning is naturally suited for use in such situations.
- *Wearing different hats*: In order to support the construction of the prospective computer science teachers' professional perception in the MTCS course, it is important that during the course, the students experience wearing different hats (see Fig. 2.1). At times, the prospective computer science teachers wear the hat of a high school pupil and are asked to perform "pupil assignments"; at other times, they wear the hat of the computer science (future) teacher; and yet at other times they wear the student's or the researcher's hats. As it turns out, active learning enables the switching between such situations in a very natural manner.
  It is also important to mention that as future computer science teachers they will have to wear different hats in their daily work (role-model, tutor, evaluator, leader, counselor, and decision maker are just a few examples) and the experience they gain in the MTCS course could help them in performing these roles.
- *Wearing the student hat*: Since the computer science material itself is usually still fresh in the student's mind, in addition to learning the content of the MTCS course itself and the construction of the professional perception as computer science teachers, the prospective computer science teachers continue, in parallel, with their mental construction of the computer science body of knowledge. From

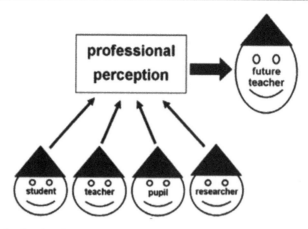

Fig. 2.1 Wearing four hats in the construction process of the prospective computer science teachers' professional perception

a constructivist perspective, in such situations, active learning is preferred over lecture-based teaching.

- *Reflection*: The prospective computer science teachers can improve the construction of their professional perception also by incorporating reflective processes into the construction process (Ragonis and Hazzan 2010). That is, by becoming reflective practitioners (Schön 1983, 1987), their comprehension of the profession of computer science education may be improved. Reflective practitioners are professionals who continuously improve their professional skills based on their on-going reflection with respect to their professional performance. Active learning is compatible with the reflective practice perspective since it provides learners with an opportunity to reflect on the activities they perform as part of their active learning.
- *Teaching methods*: Active learning enables the illustration of different teaching methods. Consequently, it enables to expose the prospective computer science teachers to different teaching methods and class arrangements. Based on the constructivist approach, the prospective computer science teachers' experience of different teaching methods in an active learning fashion, promotes their understanding of the methods' advantages and disadvantages.
- *Bridging gaps*: Active learning can bridge gaps in the teaching experience and computer science background that exist among the students participating in the MTCS course. Some of them may have stronger backgrounds in computer science; others may have more teaching experience. Since active learning enables each student to continue with the construction of his or her professional perception from his or her current professional stage, active learning can help instructors of MTCS courses overcome these variations that exist among the students.
- *High-order thinking tasks*: Last, but not least, active learning enables to offer the prospective computer science teachers tasks that enhance higher-order thinking, such as analysis, synthesis, and evaluation tasks.

2.4   Active-Learning-Based Teaching Model

So far, we have explained the rationale for the implementation of an active-learning-based teaching approach in the MTCS course. We now propose a model for active-learning-based teaching to be employed when teaching MTCS courses. This model is used in most of the activities presented in this Guide.

The active-learning-based teaching model consists of four stages—trigger, activity, discussion, and summary, focusing on a particular topic addressed in the MTCS course. The model is illustrated in Fig. 2.2 and is described in what follows. Needless to say that, when appropriate, (a) this model can be implemented in additional computer science teaching situations and (b) variations in the model implementation can be made.

First Stage: Trigger. Following the constructivist perspective, the objective of this stage is to introduce a topic with a worthwhile assignment in a nontraditional fashion (Brooks and Brooks 1999). For this purpose, the prospective computer science teachers are presented by a challenging active-learning-based trigger, an open-end activity of a kind with which they are not familiar. Specifically, a trigger should enhance and foster meaningful learning and should have the potential to raise a wide array of questions, dilemmas, attitudes, and perceptions. Following Newman et al. (2003), it is proposed that a trigger should be realistically complex and relevant for the learners. Depending on the trigger's main objective, the activity can be worked on individually, in pairs or in small groups.

In the MTCS course, a trigger can be based on different kinds of activities, such as *analyzing* a class situation, *debugging* a given computer program, *composing* a test on a specific computer science topic, *designing* an exhibition poster about a particular computer science concept, *following* a visualization or animation display for a given computer program, and so on.

One of the main objectives of introducing a new topic using a trigger is to train the prospective computer science teachers how to face and deal with open-ended and unfamiliar situations. Such situations, which are so predominant in teaching in

Fig. 2.2 Active-learning-based teaching model

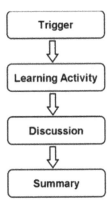

general, and in computer science education in particular, require teachers to consider multiple reaction options. In order to achieve this objective, it must be possible to approach a trigger in more than one way. Furthermore, a well-designed trigger exposes the students, while working on the trigger itself, to a rich and varied mix of computer science and pedagogical aspects. Throughout the model stages, this vast collection of ideas is discussed, elaborated, refined, and organized.

Second Stage: Activity. In this stage, the students work on the trigger presented to them. This stage may be short, or it may be longer and take up the majority of the lesson. The specific period of time dedicated to this stage naturally depends on the kind of trigger used and on its educational objectives.

Third Stage: Discussion. After the required period of time, during which the students work on the trigger either individually, in pairs, or in small groups, the entire class is gathered. At this stage, products, topics and thoughts that originated during the activity stage are presented to the entire class and are discussed. At this stage, the students refine their understanding of concepts, attitudes, and ideas, as part of the construction process of their professional perception.

The instructor highlights important ideas presented by the students and emphasizes principles derived from these ideas. In order to convey the notion that no unique solution exists for most teaching situations in general, and for the specific activity presented by the trigger in particular, the instructor does not judge students' positions and opinions. At the same time, however, classmates *are* encouraged to react and express their opinions and their constructive criticism with respect to the different ideas or materials presented.

Fourth Stage: Summary. This stage of the model puts the topic into the context of the course and emphasizes the concepts that were discussed. It is managed differently than the three previous stages. First, it is significantly shorter. Second, while in the first three stages the students are the main actors, in the Summary stage, the MTCS course instructor takes front stage. The instructor wraps up, summarizes and highlights central concepts, teaching ideas, conceptual frameworks, and other related topics that were raised and discussed during the previous three stages.

The summary can be expressed in different forms, such as a framework formulation, listing connections between the said topic and other topics, concept map, and so on.

2.5 The Role of the Instructor in the Active-Learning-Based Teaching Model

The term "instructor" refers to the lecturer teaching the MTCS course. In what follows, we explain the significant role of the instructor during each stage of the proposed teaching model.

In general, the instructor has to create a supportive intellectual and emotional environment that encourages students to be fully active.

In the first stage (Trigger), the instructor constructs and presents the trigger. As mentioned earlier, a trigger must be designed very carefully as it constitutes the basis for the entire model.

In the second stage (Activity), the instructor circulates between the different groups working on the trigger, listens to their opinions, is sensitive to what they say, and encourages them to deepen their thinking. When needed, the instructor guides the students in their discussion. Though the guidance should encourage alternative thinking approaches, the instructor is advised not to dictate any position.

In the third stage (Discussion), the instructor must act as a good listener and be sensitive to crucial points suggested by the students. Specifically, the instructor should encourage the students to explain why and how they developed their suggestions, suggest exploring different options, foster reflection processes, all without passing judgment on the students' opinions. Since well-designed triggers lead to rich discussions and debates, instructors may, at this stage, find themselves navigating through various disagreements. When needed, the instructor highlights the important facets of each opinion and presents possible connections between different ideas.

In the fourth stage (Summary), the instructor sums up the ideas presented during the previous stages. This summary is organized logically so as to highlight the main messages that were raised and discussed during the lesson. When needed, the instructor adds ideas and clarifications that were not suggested by the students themselves.

As mentioned before, the active-learning-based teaching model is used throughout this Guide in many opportunities to support the construction process of the prospective computer science teachers' professional conception as computer science teachers.

References

Anderson R, Anderson R, Davis KM et al (2007) Supporting active learning and example based instruction with classroom technology. SIGCSE'07, Covington, Kentucky, USA, pp 69–73

Ben Ari M (2001) Constructivism in computer science education. J Comput Math Sci Teach 20(1):45–73

Brooks MG, Brooks J (1999) The courage to be constructivist. Educ Leadersh 57(3):18–24

Confrey J (1995) A theory of intellectual development. Learn Math 15(2):36–45

Davis RB, Maher CA, Noddings N (eds) (1990) Constructivist views on the teaching and learning of mathematics. J Res in Math Educ. Monograph 4, Reston, VA: The National Council of Teachers of Mathematics, Inc

Gehringer EF, Miller CS (2009) Student-generated active-learning exercises. SIGCSE'09, 3–7 March 2009, Chattanooga, Tennessee, USA, pp 81–85

Hazzan O, Lapidot T (2004) Construction of a professional perception in the "Methods of Teaching Computer Science" course. Inroads—SIGCSE Bull 36(2):57–61

Kilpatrick J (1987) What constructivism might be in mathematics education. In: Bergeron JC, Herscovics N, Kieran C (eds) Proceedings of the 11th International Conference for the Psychology of Mathematics Education (PME11), vol I, pp 3–27

Ludi S (2005) Active-learning activities that introduce students to software engineering fundamentals. ITiCSE'05, Monte de Caparica, Portugal, pp 128–132

McConnell JJ (1996) Active learning and its use in computer science. SIGCSE Bull 28:52–54

McConnell JJ (2005) Active and cooperative learning: tips and tricks (Part I). Inroads—SIGCSE Bull 37(2):27–30

Newman I, Daniels M, Faulkner X (2003) Open ended group projects a 'Tool' for more effective teaching. Proceedings Australasian Computing Education Conference (ACE2003), Australian Computer Society, Inc, Adelaide, Australia

Piaget J (1977) Problems of equilibration. In: Appel MH, Goldberg LS (eds) Topics in cognitive development, vol 1. Equilibration: theory, research and application. Plenum Press, New York, pp 3–13

Ragonis N, Hazzan O (2010) A reflective practitioner's perspective on computer science teacher preparation. ISSEP2010, Zurich, Switzerland: 89–105. http://www.issep2010.org/proceedings_of_short_communications.pdf. Accessed 3 Sept 2010

Schön DA (1983) The reflective practitioner. BasicBooks, New York

Schön DA (1987) Educating the reflective practitioner: towards a new design for teaching and learning in the profession. Jossey-Bass, San Francisco

Silberman M (1996) Active learning: 101 strategies to teach any subject. Pearson Higher Education, Boston

Smith JP, diSessa AA, Roschelle J (1993) Misconceptions reconceived: a constructivist analysis of knowledge in transition. J Learn Sci 3:115–163

Walker HM (2011) A lab-based approach for introductory computing that emphasizes collaboration. Proceedings Computer Science Education Research Conference (CSERC'11), 7–8 April 2011, Heerlen, Netherlands, pp 21–31

Whittington KJ (2004) Infusing active learning into introductory programming courses. JCSC 19(5):249–259

Zimundi E (2012) Active learning for problem solving in programming in a computer studies method course. Acad Res Int 3(2):284–292

Overview of the Discipline of Computer Science

Abstract

This chapter proposes how to address in the Methods of Teaching Computer Science (MTCS) course topics associated with the nature of the discipline of computer science and with cross-curriculum topics. The importance of these topics is explained by the fact that even today, no consensus has been reached with respect to one agreed-upon definition for computer science, and different scholars view it differently. Specifically, the following topics are discussed in this chapter: what is computer science, the history of computer science, computer scientists, social issues of computer science, programming paradigms, computer science soft ideas, computer science as an evolving discipline, and computer science as an integrated and integral part of other disciplines. For each topic, its meaning and its importance and relevance in the context of computer science education are explained, and then, activities which deal with the said topic are presented.

3.1 Introduction

This chapter deals with the nature of the discipline of computer science. The importance of this topic is attributed to the fact that even today no consensus has been reached with respect to one agreed upon definition for computer science, and different scholars view it differently (see Sect. 3.2). In addition to the mere definition of the discipline, this chapter deals with additional aspects of computer science that each computer science educator should be aware of, such as the history of computer science, the scientists who shaped the field, its social aspects, and its positions within other disciplines.

This chapter also deals with two additional subjects: programming paradigms and computer science soft ideas. These topics fit into this chapter, whose aim is

© Springer-Verlag London Limited 2014
O. Hazzan et al., *Guide to Teaching Computer Science*,
DOI 10.1007/978-1-4471-6630-6_3

to provide an overview of the discipline of computer science, for several reasons. First, they are cross-curriculum topics which may foster one's attempts to acquire wider understanding of the discipline; second, these two ideas shaped major parts of the history of computer science; and third, these concepts provide cognitive tools to think with, which are considered as part of computer science.

This chapter further elaborates on the following topics:

- What is computer science? (Sect. 3.2)
- History of computer science (Sect. 3.3)
- Computer scientists (Sect. 3.4)
- Social issues of computer science (Sect. 3.5)
- Programming paradigms (Sect. 3.6)
- Computer science soft ideas (Sect. 3.7)
- Computer science as an evolving discipline (Sect. 3.8)
- Computer science—an integrated and integral part of other disciplines (Sect. 3.9)

For each topic, we first explain its meaning and its importance and relevance in the context of computer science education, and then, suggest several activities which deal with the said topic that can be facilitated in the Methods of Teaching Computer Science (MTCS) course.

The different topics and activities that the chapter deals with should not necessarily be addressed in sequential lessons and can be spread over the MTCS course. We recommend that each instructor incorporate these topics in the course when he or she feels that it is important and relevant to highlight the said topic in order to provide a broad perspective of the discipline of computer science. Such a decision may be based on different considerations. First, the instructor may notice that students' image of the discipline is too narrow (e.g., they are either not aware of the history of computer science or overemphasize programming-oriented aspects); second, when a specific computer science topic is addressed in the MTCS course, and it is relevant to associate it with one of the above topics (e.g., when the teaching of Turing machine or Dijkstra's Algorithm are discussed, it is relevant to mention the history of the discipline, in general, and the pioneers who shaped it, whose names appear in these concepts, in particular); and finally, when, from a pedagogical perspective, the instructor finds it relevant to diversify the course content by addressing topics related to the nature of the discipline rather than its core scientific topics.

3.2   What Is Computer Science?

The question "what is computer science?" does not have a single answer; different scholars emphasize different aspects of the field. Nevertheless, there is an agreement that computer science is a multifaceted field that encompasses scientific and engineering aspects, which are manifested in algorithmic problem-solving processes, for which computational thinking skills (Wing 2006), and sometimes also artistic and creative thinking, are required. It is also widely agreed that computer science is about conceptual ideas, whereas the computer serve as a means or a tool for solving computer science problems. See also, for example, Denning (2005) for different perspectives of computer science.

One important resource for the discussion about the nature of computer science is the 1989 *Computing As a Discipline* report of the ACM Task Force (Denning et al. 1989). The task force declares that "the three major paradigms, or cultural styles, by which we approach our work, provide a context for our definition of the discipline of computing" (p. 10). Specifically, they present the following three paradigms on which computer science is based: theory, rooted in mathematics; abstraction (modeling), rooted in the experimental scientific method; and design, rooted in engineering. Another source for the discussion about the definition of computer science is the lectures given by winners of the Turing Award, who, at least in the first 20 years of the discipline addressed this theme (Ashenhurst and Graham, 1987). We elaborate on the Turing Award in Sect. 3.4.

A related interesting point to address is the name of the discipline. First, as we just noted, computer science is not a pure science. Further, some computer scientists claim that it is not a science at all. For example, Abelson et al. (1996) say that "Computer science is not a science, and its ultimate significance has little to do with computers."[1] Second, the term computer, which appears in the name of the discipline, is misleading. In this context, it is relevant to quote Edsger Dijkstra, one of the pioneers of the field who said that "Computer science is no more about computers than astronomy is about telescopes."

From a pedagogical perspective, according to Ragonis (2009), an examination of the high school computer science curricula, implemented in different countries throughout the world, shows a lack of uniformity in the interpretation various bodies and countries give to curricula in computer science. The different approaches even appear in the different names of the curricula, such as information technology, information and communication technology, information systems, computer science, informatics, computer engineering, and software engineering. Sometimes, the differences between the approaches imply significant differences in the high school curricula.

Clearly, each teacher should be familiar with both the contents of the filed he or she is teaching and the nature of the field. It is suggested that this claim is amplified in the context of computer science education due to the fact different computer scientists conceive it differently. Activities 1 and 2 aim at increasing the prospective computer science teachers' awareness to these issues.

Activity 1: The Nature of Computer Science
This activity is based on five stages in which the students explore the nature of the discipline of computer science.
- Stage A: Explain what computer science is, work in pairs
  The students are asked to explain what computer science is to someone who is neither a computer scientist nor a computer science student. It can

---

[1] See also Hal Abelson's 10-min talk on What is "Computer Science"? at http://www.youtube.com/watch?v=zQLUPjefuWA.

be requested to limit the explanation to one or two sentences. From a pedagogical perspective, since the need to formulate such a brief definition for
a concept like computer science requires deep understanding of the said
topic, it is assumed that such a task fosters students' thinking about the
essence of computer science.
- Stage B: Class discussion
  The explanations proposed by the students are presented in the class. For
  each description, it is discussed what aspects of the field it highlights.
  Another option is to gather the definitions and to ask the students to categorize them according to some criterion. See Chap. 7 for additional information about classification tasks.
- Stage C: Internet exploration of computer science definitions
  Students are asked to explore (mainly by the Internet) different definitions
  for computer science as well as disagreements related to the nature of the
  field.
- Stage D: Summary and class discussion (can be carried out whether Stage
  C takes place or not)
  The instructor, together with the students, summarizes the different per-
  spectives of computer science, presented so far in the lesson. In this sum-
  mary, it is important to emphasize that:
  - Computer science deals with what is computable and with the charac-
    terization of these computations.
  - Computer science is a multifaceted field and is inspired by mathemat-
    ics, science, and engineering.
  - Computer science has many subfields and has interconnections to other
    disciplines, such as biology, economics, medicine, and entertainment.
  - The name of the discipline is misleading, and sometimes there is a ten-
    dency to confuse it with computer applications.
  The above characteristics of the field of computer science set special chal-
  lenges for computer science educators. With this respect, several questions
  can be asked and discussed in the MTCS course:
  - Should the teaching of computer science be different from the teaching
    of math, science, engineering, and art? If yes—how; if not—why?
  - Taking into consideration the above characteristics of computer science:
    - Is it important at all to teach computer science in the high school?
    - How should the first lesson in a high school computer science class
      be planned?
- Stage E: Review the Computing Curricula, homework
  Review the 3 computer science volume of the series, developed by The
  Joint Task Force on Computing Curricula of the IEEE Computer Society
  and the Association 2 for Computing Machinery, published in 2001, 2008,
  and 2013.[2]

---

[2] See 2001: http://www.acm.org/education/curric_vols/cc2001.pdf

2008: http://www.acm.org/education/curricula/ComputerScience2008.pdf
2013: http://www.acm.org/education/CS2013-final-report.pdf.

> What do these reports tell us about the connections between the nature of computer science and its teaching?

Another (or additional) task can ask the students to add/change the definition for computer science presented in Wikipedia (and, if relevant, the definition of other computer science concepts).

Activity 2: Computer Science and Other Sciences

This activity is related to the comment stated above with respect to interconnections of computer science with other disciplines, such as biology, economics, medicine, and entertainment.

- Stage A: Connections between computer science and other sciences, individual/team work
  The students are asked to identify a science which is interconnected somehow to computer science and to define the connection between the two sciences. This work can be based on Internet resources and journals.
  One recommended resource for this task is the Microsoft Corporation's report *Towards 2020Science*, published in 2006 (Microsoft Research 2006).[3] According to that Web site, the report "sets out the challenges and opportunities arising from the increasing synthesis of *computing and* the sciences."
- Stage B: Presentations
  The students present, in front of the class, the science they focused on as well as its interrelationships to computer science. It is recommended to upload the students' products to the course Web site.
- Stage C: Class discussion
  The discussion that follows these presentations has the potential to deepen students' understanding with respect to what computer science is on the one hand, and on the other hand, to highlight how computer science is used by and connected to other sciences.

For illustration, we examine the case of computer science—biology interrelation. According to the *Towards 2020Science* report (2006), "computer science is poised to become as fundamental to biology as mathematics has become to physics." (p. 8) Specifically, "one of the first glimpses of the potential of computer science concepts and tools, augmented with computing, has already been demonstrated in the Human Genome Project, and by the success of structural biology to routinely decipher the three-dimensional structure of proteins. In this and in related sequencing projects, scientists use computers and computerized DNA sequence databases to share, compare, criticize, and correct scientific knowledge, thus converging on a consensus sequence quickly and efficiently." (p. 24). Further, the report states that "there is a

---

[3] The report is available online at http://research.microsoft.com/towards2020science/background_overview.htm.

growing awareness among biologists that to understand cells and cellular systems requires viewing them as information processing systems, as evidenced
by the fundamental similarity between molecular machines of the living cell
and computational automata, and by the natural fit between computer process
algebras and biological signaling and between computational logical circuits
and regulatory systems in the cell." (p. 8).

3.3 History of Computer Science

By dealing with the history of computer science in the MTCS course, four main
objectives may be achieved.[4] First, the lesson may elevate the students' awareness
to the very existence of the history of computer science and to its main milestones.
Second, such a discussion may contribute to the professional perception of the prospective computer science teachers (Hazzan and Lapidot 2004). Third, the lesson
may increase students' awareness to the fact that computer science is beyond programming. And finally, the discussion about the history of computer science provides the students with ideas for beyond-the-curriculum activities to be facilitated
with their future pupils, such as personal projects, class presentations, and whole-
class project, in which the pupils construct the timeline of computer science when
each pupil contributes one milestone on this timeline. It is important to note that
such a project not only contributes to pupils' knowledge about computer science,
but also enhances their creativity, curiosity, and collaboration with their peers.

Activities 3 and 4 focus on the history of computer science.

Activity 3: Plan a Lesson About the History of Computer Science
This activity describes a lesson of an MTCS course we taught that focuses on
the history of computer science.
- Stage A: Introductory questions
  The lesson starts by asking the students what they know about the history
  of computer science. Usually, the proposed answers can be summarized in
  a single sentence: "It is short; about 60 years."
  The next question to be presented is: "Why is it important that computer
  science teachers know the history of computer science?" Typical answers
  suggested at this stage are: "To exhibit their knowledge to their pupils,"
  "To have a wide perspective about what they teach," "To understand the
  curriculum" and "To become familiar with people who influenced the
  development of the field."
- Stage B: Plan a lesson, work in pairs
  At this stage, when the students are conceived that as computer science
  teachers they should know more about the history of computer science,
  they are presented with the following task: Plan a lesson about any com-

---

[4] Based on Hazzan and Lapidot (2006).

puter science topic that includes a 10-min presentation about the history of computer science.

Many online resources are available on this topic. One of them is the Computing Research Association's report on *Using History To Teach Computer Science and Related Disciplines.*[5]

- Stage C: Presentations

  The prospective computer science teachers present their 10-min presentation in front of the class about the history of computer science.

  For illustration, we present two topics chosen by students in our MTCS course and their contributions to the students' perspective on the discipline.

  - *The history of object-oriented software development:* Specifically, *different programming* languages that marked the development of this paradigm (e.g., Simula, C, C++, Java) were mentioned. Furthermore, the students' realization that the concept of object-oriented development was first introduced in the early 1960s, highlighted the fact that it sometimes takes time for new ideas to be accepted by a community of professionals. This presentation formed a basis for a discussion about questions, such as: Why was a new paradigm needed? Why were several programming languages developed for the implementation of the object-oriented paradigm?

  - *The history of computers*: The fact that the first computing machines were developed about 200 years ago (and even earlier) led to a sequence of discussions. One of them addressed the question: "When did such machines start being computers?" That discussion was followed by a discussion about the definition of the term computer. This presentation formed a basis for a discussion about topics related to the teaching of computational models (e.g., Turing Machine) that, in fact, have not been actually constructed, and about connections between the conceptual and technological developments of computer science.

- Stage D: Class discussion

  After the prospective computer science teachers present their presentations, the question that opened the lesson—What do you know about the history of computer science?—is re-examined. Naturally, by this time, the one-line answer given at the beginning of the lesson ("It is short; about 60 years") is widely and deeply enriched. The lesson ends with a discussion of whether the history of computer science is really that short.

  This lesson can be continued in different ways, including a task that delves into some of the topics discussed in the lesson, a lesson about computer scientists (see Sect. 3.4), or another lesson that explores how to integrate the history of computer science into the high school computer science curriculum.

---

[5] See http://www.cra.org/uploads/documents/resources/workforce_history_reports/using.history.pdf.

Activity 4: History of Computational Machines
This activity focuses on the first computers. The students are asked to work in teams and to look for old computation machines. After the results of this searching process is briefly summarized in the class, each group explores one model/machine, addressing how it worked, how it fitted into the specific historical point of the development of computer science, and the computer scientist(s) that was/were involved it and contributed to its development. This task can be connected naturally to the topic of computer scientists (presented next). In addition, it can lead to a discussion about the parallel, yet inter-twined, developments of the technological facet of computer science and its conceptual facet, and the interactions between the two, when one facet pre-cedes the other and pushes the development of both facets.

3.4 Computer Scientists

The topic of computer scientists is a central part of the history of computer science that computer science teachers should be familiar with. Their familiarity with the topic may enable them first, to present to their future pupils some computer science topics more vividly by connecting them to peoples' stories, and second, to improve their own understanding of the development of computer science.

We suggest two options how to teach this topic. Some educators prefer to present computer scientists with relation to a specific computer science topic; others prefer to dedicate a lesson to the people who shaped the discipline in order to highlight its human aspect. In any case, it is recommended to let the students be aware of the fact that they are familiar with several computer scientists from their computer science studies so far. For example, they probably have heard about Edsger Dijkstra (from their studies of graph algorithms) and about Alan Turing (when learning Turing Machine in the Computability course or the Turing test as part of the Artificial Inelegance course).

In order to increase the students' awareness to this aspect of the history of computer science, it is recommended first, to discuss in the MTCS course the two alternatives for teaching this topic and second, to dedicate one lesson of the MTCS course for students' short presentations (15 min each) on one computer scientist.

One source to look for a computer scientist is the list of the Turing Award recipients. The award is named after Alan Turing, who has already been mentioned several times in this chapter. He was a British mathematician and one of the computer science pioneers. The A. M. Turing Award is given annually by the Association for Computing Machinery to "an individual selected for contributions of a technical nature made to the computing community. The contributions should be of lasting and major technical importance to the computer field."[6] Needless to say that this list does not exhaust the list of computer scientists, and other computer scientists as well contributed significantly to the field.

---

[6] Source: http://en.wikipedia.org/wiki/Turing_Award#cite_note-ACM-1;http://awards.acm.org/homepage.cfm?srt=all&awd=140.

The homework presented in Activity 5 guides the students how to prepare these presentations.

Activity 5: Preparation of a Presentation on a Computer Scientist, Homework
Select a computer scientist and prepare a short presentation (15 min) about this scientist. In your presentation:
1. Sketch briefly the scientist's biography.
2. Describe his or her major contributions to the field of computer science.
3. Focus on one problem that the scientist worked on and explain its impact on the field of computer science.

3.5 Social Issues of Computer Science

This section illustrates the actual teaching in the MTCS course of ethics, diversity, and soft skills—social issues of computer science which deal with the community of computer science professionals.[7]

The need to tackle such topics in the framework of computer science teacher preparation programs emerges from the growing recognition that such topics are indeed part of the discipline of computer science. This fact is well reflected, for example, in the Computer Science volume of the Computing Curricula 2001, 2008 and 2013.[8] According to these volumes, one body of knowledge of computer science is SP—Social and Professional Issues—which is composed of 11–16 core hours and includes the following core topics: Social Context, Analytical Tools, Professional Ethics, Intellectual Property, Privacy and Civil Liberties, Professional Communication, Sustainability, History, Economies of Computing, Security Policies, and Laws and Computer Crimes (taken from the 2013 curriculum).

We would add that the attention given to social issues is highlighted also in the context of other educational domains. For example, this perspective is integrated in the Science, Technology, and Society (STS) movement that "studies the relationships between these three elements and combines a cross-disciplinary approach of engineering, humanities, natural sciences, and social sciences."[9]

There are additional social issues related to computer science, such as, program comprehension, software requirements, business issues, and more (cf. Tomayko and Hazzan 2004). In what follows, as mentioned, we illustrate the teaching of social issues of computer science in the MTCS course by focusing on ethics, diversity and soft skills.

---

[7] Based on Hazzan and Lapidot (2006).

[8] See Computing Curricula:
2001: http://www.acm.org/education/curric_vols/cc2001.pdf
2008: http://www.acm.org/education/curricula/ComputerScience2008.pdf
2013: http://www.acm.org/education/CS2013-final-report.pdf.

[9] Source: http://en.wikipedia.org/wiki/Science,_technology_and_society.

3.5.1   Ethics in Computer Science Education

Ethics is part of the discipline of philosophy. The New Shorter Oxford English Dictionary defines ethics as "the science (or set) of moral principles; the branch of knowledge that deals with the principles of human duty or the logic of moral discourse." The Webster's Collegiate Dictionary adds that ethics is "the discipline dealing with what is good and bad and with moral duty and obligation."

Some communities of practice have a well-defined code of ethics (e.g., The Code of Medical Ethics). The role of such codes of ethics is to guide professionals how to behave in vague situations where it is not clear what is right and what is wrong. The need for a code of ethics arises from the fact that any profession generates situations that can neither be predicted nor answered uniformly by all members of the relevant professional community. In practice, ethics is most often needed when a conflict arises, between two (or more) possible legal actions. Since all of the alternatives are legal, ethics may help solve conflict of interests, at least in part.

A relevant question to be asked at this stage is: Does the community of computer science educators need a code of ethics? If yes, what situations should be addressed by such a code of ethics? What should be its principles? Clearly, it is not our intention in this lesson of the MTCS course to formulate a code of ethics for the community of computer science educators. However, since there are cases in which the ethical dilemmas faced by computer science educators are similar to those faced by computer scientists, and since, in addition, there are situations unique to computer science teachers, the ethics of computer science educators may be derived both from educational ethical norms and the ethical norms of the community of computer scientists. In other words, in practice, computer science educators should base their ethical norms on one of the many available educational codes of ethics and on the Software Engineering Code of Ethics and Professional Practice,[10] formulated by an IEEE-CS/ACM Joint Task Force, which outlines how software developers should adhere to ethical behavior.

This perspective highlights the ethical complexity that computer science educators must deal with as well as the importance of dealing with the concept of ethics within the framework of computer science teacher preparation.

Activity 6 aims at increasing the students' attention to situations that their profession—that is, computer science teaching—might bring them face with, and at delivering the message that their behavior should be based on ethical norms.

> **Activity 6: Analysis of Ethical Dilemmas**
> - Stage A: Case analysis, group work
>   The following two cases are presented to the students. They present a hypothetical situation that raises an ethical dilemma: the first relates directly to a school situation; the second—indirectly. Each group discusses whether or not there is an ethical dilemma in the described situation, and, if so, the students are asked to identify and describe it.

---

[10] See http://www.acm.org/constitution/code.html.

- *Case 1*
  One of your pupils downloads a software tool from the Web that helps your students in their understanding of one of the complicated topics studied in the high school computer science curriculum. The rules state that the software should be paid for after 30 days of trial usage. As it turns out, your school does not have the required budget to pay for it. It is, however, possible to reinstall the software in 30-day cycles and avoid payment.
  - How can the ethical dilemma (if exists) be solved?
  - How would you behave in such a situation?
  - Formulate (one or more) ethical principles that will help you make a decision in similar cases.
- *Case 2* (based on Tomayko and Hazzan 2004)
  Your friend works for a software house that specializes in the development of computer games. Recently, several publications have indicated that these games influence some children negatively. These games, however, are the main products of your friend's company and without them the company may not be able to survive. The company's management is aware of these publications and gathers all the employees to discuss the future of the company. You are invited to the meeting as a representative of the educational system.
  - Suggest different opinions that might be expressed in the meeting. What (if at all) ethical consideration does each of them represent?
  - What (if at all) conflicts of interest are presented in this case?
  - What is your opinion with respect to this case?
  - How would you behave in such a case?
  - Formulate (one or more) ethical principles that might help make a decision in similar cases.
- Stage B: Presentations and discussion
  After the students work on the task, the different groups present their works and express their opinions with respect to the different questions. Specifically, after the source of ethical dilemma is observed in each case, the focus is placed on specific solutions that might solve it. This discussion delivers the message that ethical considerations should be part of the profession of computer science teaching and that different solutions can be found that eliminate the ethical dilemmas (at least partially).

3.5.2 Diversity

Diversity is expressed in different ways in any community, and, in particular in any computer science class that the prospective computer science teachers may teach in their future professional life. Diversity can be exhibited, for instance, in terms of students' fields of interest, ways of thinking and perspectives, background, gender, nationality, and more. Accordingly, prospective computer science teachers should be familiar with the concept of diversity in general and, in particular, with its importance in the context of computer science education and with how to take advantage of diversity in their teaching processes.

Activities 7, 8, and 9 address diversity. Activity 7 deals with diversity in general, Activity 8—with different approaches toward students' performance, and Activity 9—with gender diversity—a well-recognized topic in computer science education (cf. e.g., the June 2002 Special Issue of *Inroads—The SIGCSE Bulletin*).

We note in passing that, in fact, the underlying idea of presenting triggers in itself represents *diversity,* since triggers should be designed in a way that encourages different perspectives and ways of thinking. Furthermore, the pedagogical viewpoint of constructivist teachers, which legitimizes and respects the differences between students, is based on diversity.

Activity 7: Diverse Class Demography, Group Work
The students are asked to describe a class demography that is diverse as much as possible. This trigger is presented to the students before the concept of diversity is introduced. After the students work on the trigger, each group presents its suggested hypothetical class in the course milieu. The collection of class structures presented at this stage by the students indicates their level of awareness to diversity, increases their awareness to diversity, and leads to discussions on how to deal with diversity. A follow up activity can focus on differences and similarities between these different class structures and their implications in computer science lessons.

Activity 8: Test Evaluation, Work in Pairs
The students are asked to evaluate several pupil answers to a question given on a test, administrated at a high school class. After working on the trigger, each group suggests its philosophy with respect to the evaluation of the answers. The different perspectives exhibited at this stage highlight that there are different perspectives regarding student errors, that this is another way by which diversity is expressed, and that different considerations play in such teaching situations.

For additional information about the theme of evaluation in the MTCS course see Chap. 10 and Lapidot and Hazzan (2003).

Activity 9: Gender Diversity, Open Conversation
Table 3.1 is presented (empty) to the students and they are asked to suggest factors that encourage each gender (or both genders) to study computer science or discourage each gender (or both) from studying computer science.

While the students suggest their ideas, the table is filled accordingly. When the students' suggestions are presented, their opinions should not be judged and the instructor should simply write what is said in the table in

Table 3.1 Factors that encourage/discourage males/females in the choice of computer science

	Boys	Girls	Both genders
Encourage	[2]		
Discourage		[1]	

the appropriate cell. Our experience says that in most cases, the following picture is clearly observed at the end of this process: Cell [1], which includes factors that *discourage girls* from studying computer science, and Cell [2], which includes factors that *encourage boys* to study computer science, are full, while the other cells remain almost empty.

The class discussion that follows this trigger leads the students to rethink their perspective regarding gender diversity in computer science classes. Particularly, the students begin observing their bias very clearly. This increased awareness is used in order to further discuss with them how to encourage diversity in their future classes.

Following this discussion, data about gender diversity (such as, statistics from different countries) are presented to the students and the importance of diversity is emphasized.

3.5.3 Soft Skills

According to Sukhoo et al. (2005), "Soft skills […] is concerned with managing and working with people, ensuring customer satisfaction with the intention of retaining them and creating a conducive environment for the project team to deliver high quality products within budget and on time and exceeding stakeholders' expectations" (p. 693–694).

A fast Google search suffices to reveal that almost all problems associated with software development processes are connected to people (mainly customers and team members) and that their origin is rooted not in technological aspects but in cognitive and social aspects—i.e., the expression of soft skills. Therefore, it is important to address the teaching of soft skills in the MTCS course.

Hazzan and Har-Shai (2013, 2014) propose teaching soft skills using active learning (see Chap. 2) by applying:

1. General teaching principles: e.g., semi-situated learning (where content is presented in an authentic context and learning is based on social interaction and collaboration; Lave and Wenger, 1991)
2. Soft skills teaching principles: e.g., teamwork, reflection, and diversity (addressed also in this book)

Accordingly, the following activities all aim to enable the prospective computer science teachers to practice these teaching methods both as learners and as teachers, by holding open discussions, giving examples of real cases, reflecting on personal experiences, and role-playing and simulating real life situations.

Activity 10: Introduction to Soft Skills, Homework Toward the Lesson on Soft Skills

1. Table 3.2 presents many soft skills. Characterize each of them and based on this characterization, define the concept of soft skills.
2. Select 2 soft skills. For each of them, suggest three activities by which you would teach them to:
   - High school students;
   - Undergraduate students.

Reflect: Are the activities for the two populations similar? Different? If they are similar—explain why? If they are different—explain how?

3. Select five soft skills that in your opinion are the most important for computer science students either in the academia or the industry. For each of them, ask your students to develop an illustrative scenario, taken from the computer science world.

Table 3.2 A list of soft skills

Flexibility and ability to compromise	Inter-personal communication skills	Ability to ask for and receive assistance
Identification and location of relevant information	Ability to connect with a wide variety of people	Openness toward and understanding of organizational changes
Ability to express oneself in a concise and matter-of-fact manner	Ability to analyze data, drawing conclusions, and acting upon them	Ability to formulate and understand requirements
Openness	Patience	Teamwork
Learning abilities	Decision making	Listening skills
Respect for diversity	Risk assessment	Working with clients
Time planning and scheduling	Teaching and instructing skills	Knowledge management
Teaching and instructing colleagues	Remedying unpleasant situations	Budget planning and staying within budget
Coping with uncertain situations	Ability to learn from mistakes	Ability to adapt to changing situations
Fund raising	Receiving feedback	Public speaking
Solving and managing conflicts	Ability to learn from mistakes	Interpersonal communication skills
Self-awareness	Interviewing skills	Ability to initiate
Team management	Identification and exploitation of professional opportunities	Creativity

Activity 11: Definitions for Soft Skills, Teamwork at the Beginning of the First Lesson on Soft Skills

The following definitions were given by undergraduate students who learned a course on soft skills, in which the concepts of soft skills were left undefined until the 11th session of the course. Yet, as illustrated below, the students had already gained some sense of the meaning of the concept, and when asked to define the concept in their last homework assignment, they did so, highlighting its multifaceted nature, as illustrated by the following examples:

- A soft skill is a personality trait, an interpersonal attribute or skill that, together with hard skills, enhances work and is manifested during interactions with people.
- Soft skills are the collection of a person's traits, communication and language capabilities, friendships, etc. that characterize the person's relations with other people.
- A soft skill is a personality trait that enhances the individual's reciprocal relations and work performance. As opposed to hard skills, which are a person's set of skills and the ability to perform a certain kind of activity, soft skills refer to the person's ability to communicate effectively with colleagues and clients. Soft skills also apply extensively both in and outside the workplace.
- Soft skills characterize the person's character, traits, emotions, creativity, etc. They are perceived as being inherited, but it turns out that they can be learned over time and with practice.
- A soft skill is a personal trait that can improve the individual's personal reactions and increase efficiency and work performance, and enhance any other situation that involves teamwork.

Discuss:
- What do these definitions reflect?
- Which definition would you prefer?
- Present your definition for the concept of soft skills.
- Explain the importance attributed to soft skills in computer science and software engineering professional work.

Activity 12: Rank Your Soft Skills, Homework, After the Lesson on Soft Skills

- Stage A: Individual work
  Choose five soft skills that in your opinion you are good at and five soft skills that in your opinion you should improve at.
  1. For each soft skill that you are good at, describe a case that illustrates this.
  2. For each soft skill that you should improve at, suggest how you can improve and how you will document the improvement process.
- Stage B: Forum discussion
  Each student describes at a form his/her product of Stage A and give feedback to other five classmates.

- Stage C: Individual reflection (submitted online)
  1. In what ways is your list of soft skills similar to those of your classmates?
  2. Suggest five habits of teaching that you will adopt as a Computer Science teacher to promote the soft skills of your prospective students.

3.6 Programming Paradigms

A paradigm is a way of doing and seeing things, a framework of thought in which one's world or reality is interpreted.[11] The concept paradigm became popular in the scientific world mainly following Kuhn's book *The Structure of Scientific Revolution* (Kuhn 1962), in which he used this term with respect to a conceptual world view that consists of formal theories, classic experiments, and trusted methods.

The topic of *programming paradigms* is a multifaceted notion and includes aspects of mental processes; problem-solving strategies; interrelations between paradigms, programming languages, programming style, and more. We note that while a programming paradigm is a heuristic for solving algorithmic problems, a programming language is a means of expression for a programming paradigm.

The concept of programming paradigm is defined in different ways (see, e.g., Abelson et al. 1996; Ambler et al. 1992; Floyd 1979; Sethi 1996; Tucker and Noonan 2002; Van Roy and Haridi 2004; Watt 1990), each emphasizing different aspects of the concept. The following definition, for example, which is composed of several of the resources mentioned above, summarizes the major points:

Programming paradigms are heuristics used for solving algorithmic problem. A programming paradigm analyzes a problem through specific lens, and based on this analysis, formulates a solution for the given problem by breaking the solution down to specific building blocks and defining relationships among them.

We mention several major programming paradigms: procedural (imperative), object-oriented, functional, logical, and concurrent.

Many researchers emphasize the importance of both teaching the subject of programming paradigms to students and exposing students to a number of programming paradigms (Carey and Shepherd 1988; Floyd 1979; Haberman and Ragonis 2010; Van Roy et al. 2003). We mention three reasons that explain the importance attributed to the learning of several programming paradigms: the development of cognitive tools, the ability to explore the same concept/problem within different paradigms improves learners' understating of the said concept/problem, and increasing learners' flexibility in problem-solving processes of different kinds of problems.

At the same time, there are programming language courses that emphasize the programming language aspect, while less attention is paid to the programming paradigm aspect. This can be explained by the fact that the concept of a programming

---

[11] Based on Stolin and Hazzan (2007).

paradigm is considered to be a soft idea (see Sect. 3.7 of this chapter), as opposed to rigid content, such as programming language syntax (Corder 1990; Turkle 1984); therefore, teaching programming paradigms is not a simple task. For these reasons, we decided to focus in this guide to the teaching of programming paradigms.

A discussion about the concept of programming paradigm in the MTCS course has several additional advantages. First, it enables the instructor of the MTCS course to discuss with the students the meaning of the levels of abstraction (a central computer science concept) within a relatively familiar context, that is, the relations between programming paradigms and programming languages. Second, since in most cases, high school computer science curricula are partially based on programming, the students' familiarity with the concept of programming paradigm, including the differences and connections between this concept and the concept of programming language, may increase their awareness in their future teaching to the difference between technical aspects of programming languages and conceptual computer science ideas. Such understanding may help the prospective computer science teachers teach computer science also on a higher level of abstraction. Third, it may improve students' understanding of the essence and history of the discipline of computer science. Finally, it is reasonable to assume that not all students have learned a course that focuses on the notion of programming paradigm; the MTCS course can be, therefore, an appropriate opportunity to partially close this gap.

Activities 13, 14, and 15 that explore the notion of programming paradigm are relatively advanced; therefore, they should be facilitated in the MTCS course only with students who have the needed background in computer science. Specifically, before the activities are facilitated, it is important to verify that the students are familiar with at least two programming paradigms; if they are not familiar with at least two programming paradigms, the instructor of the MTCS course may consider dedicating 1–2 lessons to the teaching of a programming paradigm with which the students are not familiar. Such lessons will provide the instructor with an additional opportunity to discuss the differences between programming paradigms with the students and to reemphasize the fact that programming languages are, in fact, an expressive mechanism for programming paradigms.

Activity 13: Programming Paradigms—Exploration of Learners' Knowledge
The aim of this activity is to let the instructor of the MTCS course observe the students' level of familiarity with the concept of programming paradigm. It should be facilitated, as mentioned above, only after the instructor verifies that the students are familiar with at least two programming paradigms, or at least with two programming languages, which represent two different programming paradigms.
- Stage A: Worksheet, individual work
  The students are asked to work individually on the worksheet presented in Table 3.3.
- Stage B: Class discussion
  After the students work on this worksheet individually, the instructor collects their answers and discusses them with the class; when needed,

Table 3.3 Worksheet on programming paradigms

Worksheet—programming paradigms
Answer the following questions
1. What is a programming *language*?
2. What is a programming *paradigm*?

3. Give examples of three programming paradigms. For each paradigm
   —explain why it is a programming paradigm
   —list at least two programming languages that implement the said paradigm

4. Compare the two concepts: a programming paradigm and a programming language

clarifications are added by the instructor. At the end of this discussion, the instructor summarizes the activity by:

- Presenting a definition for the concept of programming paradigm
- Presenting the differences between several programming paradigms (according to students' knowledge)
- Highlighting the difference between a programming paradigm and a programming language and illustrate them with several examples
- Explaining that since programming languages implement programming paradigms, different programming languages may implement the same programming paradigm; nevertheless, if one is familiar with one programming language which represents a specific programming paradigm, it is reasonable to assume that he or she will be able to switch smoothly to another programming language that represents the same programming paradigm
- Highlighting the fact that the differences between programming paradigms are more fundamental and therefore, unlike switching between programming languages that represent the same programming paradigm, the switching between programming paradigms is not (in most cases) a trivial cognitive task.

As mentioned, Activity 13 allows the instructor of the MTCS course to identify students' current knowledge with respect to the notion of programming paradigms. Based on this observation, the instructor can move on and continue the lesson with either Activity 14 or Activity 15 or both.

Activity 14: Abstract-Oriented Examination of Programming Paradigms

- Stage A: Problem solving in different programming paradigms, work in pairs
  The students are asked to work in pairs on the worksheet presented in Table 3.4. Clearly, another problem can be presented in the worksheet, for example, one of the sort algorithms.

Table 3.4 Problem solving in two programming paradigms

Worksheet—problem solving in different programming paradigms
PartA
Solve the following task in at least two programming paradigms. Do not implement the solution. Given an arrangement of domino stones, determine whether or not it is a legal arrangement and return True or False accordingly

For example, this is an illegal arrangement of Domino stones:	2 3 | 4 2 | 2 2 | 3 5
and this is a legal arrangement of Domino stones:	2 3 | 3 5 | 5 4 | 4 5

PartB
After solving the task, reflect
What stages you went through while working on the solution in each programming paradigm? Were the stages similar? Were they different?
For each programming paradigm, describe which of its characteristics are expressed in your solution.
If you had to choose a programming paradigm to solve the problem, what would be your preferable paradigm? Explain your choice

Part A requires students to think about the level of abstraction represented by the programming paradigms; that is, to think on a relatively higher level of abstraction. Part B, in which they are asked to reflect on the strategies they employed in Part A, further increases the abstraction level of students' thinking since they are asked to reflect on their own problem-solving strategies and to discuss similarities and differences between programming paradigms.

- Stage B: Class discussion
  The class discussion that summarizes the activity should highlight the fact that programming paradigms provide a context in which a discussion on different levels of abstraction can take place (e.g., by addressing mechanisms that programming paradigms and programming languages provide to express and create abstraction).

Activity 15: Activity Design for a Given Programming Paradigm

Based on the above activities, in this activity the students practice the construction of tasks that are suitable to be solved by a specific programming paradigm. The importance of this activity derives from the fact that the curriculum the prospective computer science teachers will teach in the future is probably based on a specific programming paradigm. Therefore, they should be aware of the fitness of the tasks that they will develop for their future pupils to the programming paradigm used by the curriculum.

- Stage A: Tasks and paradigms, group work
  The students are asked to work in groups, to choose two programming paradigms with which they are familiar, and for each paradigm:

> - Bring an example for a task that is suitable to be solved by the said
>   paradigm.
> - Explain why the task is suitable to be solved by the said paradigm.
> - Solve the task with the said programming paradigm.
> - Stage B: Reviewing the groups' work
>   Each group presents its tasks and explains its considerations in the design
>   process of the tasks for each paradigm. For each presented task, the other
>   class members are asked to express their opinion whether or not the task
>   fulfills the requirements, that is, whether the task fits to be solved by the
>   proposed paradigm.
>   For selected tasks, the instructor asks whether it can be naturally solved by
>   another programming paradigm than the one proposed by the team which
>   presented it. If it is, the students can be asked to explain why and to solve
>   it in the other paradigm; if it is not—the instructor can ask what changes
>   should be made in the question formulation to make it naturally solvable
>   also by applying another paradigm. The changes suggested by the students
>   can then be analyzed and conclusions with respect to the kind of changes
>   can be fostered.
> - Stage C: Class discussion
>   The summary should address the different considerations addressed by the
>   groups while designing questions for different programming paradigms. It
>   is relevant to raise the question whether similar considerations were used
>   for all paradigms. Such a discussion, once again, enables to highlight first,
>   the different perspectives that different programming paradigms exhibit in
>   problem-solving situations and second, the variety of the levels of abstrac-
>   tion on which the topic of programming paradigms can be thought of.

3.7   Computer Science Soft Ideas

In this section, we first explain the essence of computer science soft ideas
(Sect. 3.7.1) and then address its position in the MTCS course (Sect. 3.7.2).

3.7.1   What Are Computer Science Soft Ideas?

This section focuses on the teaching of computer science soft ideas.[12] According
to Hazzan (2008), a soft idea is a concept that can be neither rigidly nor formally
defined, nor is it possible to guide students as to its precise application. Indeed, how
can stepwise refinement be presented formally? How can one outline a list of "how
to" actions for seeking a good algorithm? How can the move between different

---

[12] Based on Hazzan (2008).

Table 3.5 Rigid and soft facets of the concept of variable

	Rigid facets	Soft facets
Variable name	Syntax rules	The need for a name
		What a meaningful name is and why it is important
Variable value	Type rules, memory allocation	A variable has a single value at any point but it can change over time
Assignment	Syntax rules	The need for assignment
		The importance of an initial assignment to variables

levels of abstraction be formally described? And how the decision of a specific data structure that fits for the solution of a given problem can be instructed? To grasp the essence of soft ideas and the way in which they differ from rigid concepts, it is sufficient to compare the answers to the above questions with answers to questions such as the following: Is it possible to formulate the syntax of a programming language? What is the formal definition of a heap? In most cases, syntax rules are considered as rigid facets while semantics rules—as soft facets.

Having said this, it should be noted that computer science concepts are neither soft nor rigid; rather, each computer science concept has some soft aspects and some rigid aspects. Table 3.5 reflects this idea with respect to the concept of variable. Nevertheless, some computer science concepts, such as the one mentioned above, lend themselves to be softer than others.

In fact, soft ideas are part of any profession, including all fields of science and engineering. In computer science, however, soft and rigid concepts are tightly connected. This connection is required mainly due to the frequent need to implement a problem solution in some programming language and due to the high cognitive complexity involved in the field. This cognitive complexity can be partially explained by the fact that in many cases the objects of thoughts are implemented by an intangible entity—that is, software. In this spirit, Dijkstra's (1986) indicates that a computer scientist should move through many levels of abstraction, starting at the level represented by the machine and ending at the level of abstraction represented by the human thinking.

Due to the special role of soft ideas in computer science, computer science professionals include the discussion of soft ideas as part of the discipline; consequently, their teaching should not be neglected.

3.7.2 Computer Science Soft Ideas in the MTCS Course

Though the centrality of soft concepts within the computer science community is highly acknowledged, due to the nature of computer science soft ideas, it is not a trivial matter to teach such concepts.

First, it is not easy to explain the essence of soft ideas, nor is it easy to explain how and when they should be approached or utilized. Unlike rigid ideas, such as some aspects of complexity and programming, soft concepts cannot be expressed formally, and additionally, it is not sufficient to present students with only the definition of the said concept. In other words, soft ideas cannot be carried out simply by applying rigid rules; in order to understand and apply a soft idea, one must sense it.

Second, in some cases, soft ideas are *not* connected exclusively to a specific topic; rather, they are expressed throughout the curriculum and, therefore, can be illustrated and applied in the context of different topics.

Third, since soft ideas are never applied out of context, their application requires awareness. More specifically, since soft ideas are usually expressed when some other topic is at the focus of the discussion, soft ideas are actually accompanying, yet important, notions that serve and support the thinking that is dedicated to the other concept. Thus, when trying to illustrate a soft idea, another topic must be taught at the same time, there must be awareness of the way in which the soft idea is thought about, and its use must be demonstrated in situations in which it can be helpful. Clearly, this is not a simple pedagogical task.

Accordingly, the motivation to include the discussion of teaching computer science soft ideas in the MTCS course results from the awareness to these difficulties together with the acknowledgment that computer science soft ideas should be addressed, at least on a basic level, in high school computer science classes.

It is appropriate to dedicate a lesson about the teaching of computer science soft ideas toward the end of the first semester of the MTCS course (either when the course extends for one or two semesters). At this stage of the course, it is reasonable to assume that the students have already heard the notion of soft ideas (as is explained below) and have enough knowledge, related to the teaching of computer science, based on which they will be able to discuss meaningfully the *teaching* of computer science soft ideas.

Having said that, it is recommended to highlight the notion of computer science soft ideas and the importance of introducing it to high school computer science pupils from the very first lessons of the MTCS course, and then, as mentioned above, to dedicate a full lesson to the teaching of computer science soft ideas toward the end of the first semester of the course. For example, in the MTCS lesson about evaluation, it makes sense that the notion of soft ideas comes up, for example, with respect to how to evaluate students' understanding of soft ideas. In such cases, it would be appropriate to mention to the students the notion of computer science soft ideas, and to specifically indicate that one of the future lessons of the course will be dedicated to their teaching. Thus, when the lesson on computer science soft ideas is taught, the prospective computer science teachers will have a basis on which to go on constructing their knowledge related to the teaching of computer science soft ideas.

From this perspective, similar to other topics, such as students' difficulties and reflection, soft ideas can be viewed as a meta-idea (or meta-theme/thread) that is intertwined in the MTCS course with respect to different topics.

Activities 16–22, to be facilitated in the MTCS course, are dedicated to computer science soft ideas. Activities 16, 17, and 18 address the notion of soft ideas in general; Activities 19–22 focus on abstraction as an example for a computer science soft idea.

As these activities illustrate, the teaching of computer science soft ideas, either in the MTCS course, in the high school, or in the university, should be done in an open, interactive, and collaborative learning environment. The activities should be open and stimulate and support learners' understanding of computer science concepts in general and computer science soft ideas in particular.

After the students work on these activities, a summary discussion can focus on additional topics, such as:

- The prospective teachers' own understanding of computer science soft ideas: What did they learn during the work on the activities? Did they improve their understanding of computer science soft ideas in general and of a specific computer science soft idea in particular? If yes—how? Do they have questions about soft ideas? If yes—what kind of questions? What is the source of these questions?
- The teaching of soft ideas in the high school: What is the uniqueness of teaching soft ideas? Is the teaching of soft ideas different or similar to the teaching of rigid computer science concepts? What challenges does the teaching of computer science soft ideas raise? How would they incorporate the teaching of soft ideas in their future teaching in the high school? Should a lesson be dedicated to soft ideas or should soft ideas be mentioned only when other topics are learned? At what stage is it appropriate to teach soft ideas to high school computer science pupils? This discussion can also address possible difficulties that the students anticipate and the difficulties that their high school computer science pupils will face with respect to the understanding of computer science soft ideas.

Activities 16–17 are triggers that can be served as an introductory task for computer science soft ideas. Since they are similar in some sense, there is no need to facilitate both of them.

Activity 16: Types of Concepts, Class Discussion
The students are asked what computer science concepts they are familiar with. While the students suggest these concepts, the instructor writes them on the board in two sets: One set consists of concepts whose soft aspects are more dominant; the other set includes concepts whose rigid aspects are more apparent. One way to determine to which set a specific concept belongs, is by checking how difficult it is either to define it formally or to present for it an algorithm to apply. For example, according to this criterion, a loop, a tree, and functional complexity would be classified as rigid concepts (a fact that does not imply, though, that they do not have soft aspects), while the concepts of algorithm, abstraction, and debugging would be classified as a soft idea. When the specific classification is not evident at a glance, for example, with respect to concepts, such as data structure or class inheritance, a third set can be created. The students should not be told how the concepts they suggest are grouped into the two (or three) sets.

When a sufficient number of concepts are presented on the board, the students are asked to suggest criterions according to which, in their opinion, the concepts were categorized. Their answers should not be judged, and they should be encouraged to suggest different ideas for the categorization criterion.

Then, the students are asked to give titles to each set that are presented at this stage on the board and to explain why they suggest the said title. Their suggestions for titles should be listed as well. It is reasonable to assume that at some stage the distinction between soft and rigid ideas will come up, even if these specific terms will not be mentioned explicitly.

When no additional suggestions for titles are offered, the instructor reviews the different titles/categorizations suggested by the students with some pedagogical comment. Then, it is declared that the lesson is dedicated to the teaching of soft ideas, and the rationale for dealing with this topic is explained (as previously described: computer science soft ideas are part of the discipline of computer science, and it is not a trivial matter to teach them).

Activity 17: Computer Science Concept Classification, Teamwork[13]
The students are asked to work in teams on the worksheet presented in Table 3.6.

After the students work on this worksheet, a discussion takes place with the whole class that aims to elicit the notion of soft ideas and its role in computer science and in computer science education.

Additional classification activities are presented in this guide with respect to control structures (in Chap. 7) and with respect to recursion (in Chap. 12).

Table 3.6 Worksheet—Computer science concept classification

Worksheet
The following list of computer science concepts is given (in an alphabetical order)
A formal language, Abstraction, Algorithm, Assignment, Branchingstatements, Class, Control structures, Correctness, Data representation, Debugging, Efficiency, Generalization, Inheritance, Input-Output instructions, Modularity, Object, Parameter, Procedure, Recursion, Sorting, Stepwise refinement, System state, Tracing, Tree, Variable
Sort the above concepts into sets
Give a title to each set
To each set, add at least one concept that does not appear in the set

---

[13] ©Migvan—Research and Development in Computer Science Education, The Department of Education in Science and Technology, Technion Israel Institute of Technology.

Activity 18: Design of Tasks and Questions About Soft Ideas

The students are asked to design an activity that deals with soft ideas. Two options are suggested: the first asks to construct a question that demonstrates one soft idea to computer science learners; the second asks to construct a question to be included in a test that checks students' understanding of one soft idea. In most cases, it is sufficient to facilitate one of these options.

Option 1: Construction of an activity that demonstrates a soft idea, work in pairs

The following task is presented to the students:

- Select one of the soft ideas presented on the board (if the first trigger was facilitated) or in the worksheet presented in Table 3.6 (if the second trigger was facilitated).
- Develop a task/question to be presented to high school computer science pupils that illustrates this soft idea.
- Explain why you selected this soft idea and what guidelines you followed when constructing the task/question.

Option 2: Construction of a question to be included in a test that checks learners' understanding of one soft idea, team work

The following task is presented to the students:

- Select one of the soft ideas presented on the board (if the first trigger was facilitated) or in the worksheet presented in Table 3.6 (if the second trigger was facilitated).
- Develop a question to be included in a test that checks students' understanding of the said soft idea.
- Describe the guidelines you followed when developing the question and formulate instructions how to check students' answer. Explain the guidelines you followed while formulating these instructions.

To further highlight the nature of computer science soft ideas, it is important to discuss the students' suggestions for these questions (see also Chap. 9 about types of questions). Further, it is important to discuss whether it is possible at all to test learners' understanding of soft ideas. Indeed, it is difficult to develop a test question that checks students' understanding of soft ideas. Therefore, for this purpose, other evaluation methods should be included in computer science education (see Chap. 10).

Activities 19–22 address computer science heuristics, such as top-down development and successive refinement, focusing on the concept of abstraction. A similar discussion, however, can be conducted with respect to other methodologies and topics, such as operating systems or data abstraction. In addition, with respect to other computer science soft ideas, it is important to note that ideas such as abstraction should not be addressed as isolated topics; rather, they should be addressed, referred to, and highlighted at any appropriate opportunity. Still, since such topics are complex in nature and are usually addressed in relation to other (sometimes complex) topics, the special attention they receive in the MTCS course can highlight their importance.

We present several tasks that can be facilitated with respect to abstraction[14]. Clearly, there is no need to facilitate all of them; each instructor should choose the activities that fit his or her class and his or her pedagogical considerations.

Activity 19: Abstraction—Definition
Based on their familiarity with the concept of abstraction from their computer science courses, students are asked to define the term "abstraction." After the definitions are collected and discussed, a short lecture is given that summarizes the students' contributions and adds concluding remarks from the computer science literature.

This activity can serve also as a good opportunity to discuss with the students the role of definitions in learning processes. For example, one teaching dilemma in this context would address the best timing to introduce a full and formal definition.

Activity 20: Abstraction—Teaching Planning
Students are asked to design activities for the teaching of abstraction. The working assumption behind this task is that when one teaches a certain concept, especially a complex concept, such as a development heuristics for computer programs, one deepens one's own understanding of that concept.

After working in teams on this task, the students present the activities to their classmates. Indeed, as is intended, when students present the activities they have designed, many issues related to the concept of abstraction are clarified, elaborated on, and refined.

Activity 21: Abstraction—Pedagogy
Students are requested to explain why methodologies, such as abstraction, are difficult to teach. In order to illustrate the pedagogical potential of this activity, we present some of the explanations presented in our classes by prospective computer science teachers:

- In order to teach abstraction, we need concrete examples, and thus we lose the generality inherent in the topic.
- Effective discussion and demonstration of the power of abstraction [...] can be carried out only when based on complex problems. These complex problems may distract our attention from the topic, and we should share our mental resources between the problem itself and thinking about the methodology (such as abstraction).

---

[14] Based on Lapidot and Hazzan (2003).

- There is a gap between programming, which is a real action, and learning problem-solving methodologies, which is about thinking.
- Since abstraction […] [is] an individual process, there is no unique way "to do" it, so how can one teach these heuristics?
- These utterances reveal that by asking the prospective teachers to analyze the teaching of a particular topic, they are indirectly induced to examine the concept itself and analyze its properties. In the above quotes, we can see how the prospective teachers refer to topics such as relationships between thinking-with-examples and abstract thinking, limitations of the human mind, thinking processes, and the fact that there is no unique way to implement ideas, such as abstraction. It seems reasonable that such analysis improves the students' understanding of the discussed topic.

Activity 22: Abstraction—Teaching Programming Heuristics
This activity consists of a class discussion on the question: Is it possible to teach programming heuristics in the same way as other computer science ideas are taught? This discussion addresses the multifaceted nature of topics, such as abstraction, and serves as an excellent opportunity to discuss the teaching of other soft ideas versus the teaching of rigid computer science concepts.

3.8 Computer Science as an Evolving Discipline

Computer science, like other disciplines, keeps changing. Nevertheless, while its core ideas remain (more or less) constant, its relevance to the society increases. In this section, we describe two topics that may reflect this idea—Big Data (Data Science) and Cyber—which we suggest to address in the MTCS course for three reasons. First, by addressing these topics in the MTCS course, the instructor can demonstrate to the computer science prospective teachers how new topics in computer science are based on its foundations; second, these topics show how computer science enhances the citizen's understanding and use of cutting edge technologies; and third, the prospective computer science teachers may improve their understanding of these topics. Indeed, in recent years, more and more courses are taught on these topics (see, e.g., Anderson et al. 2014), which reflect the more and more central position that this topic gets.

Activity 23: Acquaintance with Data Science
- Stage A: Trigger, Group work
  The aim of the trigger is to let the prospective computer science teachers experience the design of a teaching unit for a basic computer science concept while connecting it to current technological development that is most probably known to the new generation of computer science students (and teachers).
  - Search a definition for Data Science. Layout 5 of its most important characteristics and principles.
  - What are the computer science concepts related to Data Science?
  - Present at least five computer science concepts on which Data Science based. Explain your selection.
  - What connections can you draw between computer science and Data Science?
  - Following a class discussion about the students' findings, the actual work on the design of the teaching unit takes place.
- Stage B: Design activity, Group work (or individual work at home)
  Based on the above trigger, select a core computer science topic and design a lesson in which the concept is introduced from the data science perspective.
- Stage C: Group presentation in class
  After the groups complete designing the lesson, they present part of the lesson to their class milieu and get feedbacks.
- Stage D: Discussion
  It is worth listing the computer science subjects and concepts addressed in this lesson and to ask the prospective computer science teachers to draw connections between them (a concept map can be one option for this task). Such a discussion may reflect, among other ideas, the centrality of computer science in the development of the leading, cutting-edged, updated technologies which shape our lives. Such a conclusion not only highlights the importance of computer science, but can also be used to highlight the responsibility of the profession of computer science to the community and the society and be connected to the concept of ethics addresses above.

A similar activity can be done with respect to cyber and other new branches in computer science[15]. One direction that can be discussed in the MTCS course is the connectivity of computer science to many other areas (as Sect. 3.9 illustrates).

---

[15] E.g., the Cybersecurity Education at UMD program at http://cyber.umd.edu/education.

3.9   Computer Science: An Integrated and Integral Part of Other Disciplines

Computer sciences become an integral part of many disciplines. This integration is expressed in two main ways: (1) use of computational power and methods of data representations, for example, in the Human Genome Project, or in computational linguistics; (2) the use computer science computational models, for example, in the design of a biological computer.

It is important to discuss the valuable part of computer science within other disciplines when exploring the significant place of computer science these days. This issue can be explored by looking for example at the Towards 2020 Sciences (Microsoft 2005) document. To address this challenge, Microsoft brought together an international expert group for a workshop to define and produce a new vision and roadmap of the evolution, challenges and potential of computer science and computing in scientific research in the next 15 years. The document sets out the challenges and opportunities arising from the increasing synthesis of computing and the sciences, e.g., biology, earth sciences and climatology, chemistry, physics, linguistics, and economics.

Activity 24: Integration of Computer Science with Other Disciplines
- Stage A: Trigger, Group work
  Locate at least five resources which address connections between computer science and another scientific field.
- Stage B: Reflection and class discussion
  Following the students work on the trigger, a class reflection can take place in which the following issues (and many others, of course) may be addressed: Among the connections you located, what was new for you? Did you change your perspective at these domains following the above examination?
- Stage C: Summary
  It is recommended to summarize this lesson with a representative list of topics that illustrate the high connectivity of computer science to other fields, e.g.: robotics, information systems, game theory, bioinformatics, and computational biology. In parallel, the mutual contribution of computer science and these fields can be addressed.
- Stage D: Homework
  Draw a concept map (Chap. 7) that summarizes the above lesson. Upload it to the course forum and discuss with your classmates the different concept maps that each of you drew.

References

Abelson H, Sussman G, Sussman J (1996) Structure and interpretation of computer programs, 2nd edn. The MIT, Cambridge

Ambler AL, Burnett MM, Zimmerman BA (1992) Operational versus definitional: a perspective on programming paradigms. Comput 25(9):28–43

Anderson P, Bowring J, McCauley R, Pothering G, Starr C (2014) An undergraduate degree in data science: curriculum and a decade of implementation experience. Proceedings of SIGCSE 2014—the 45th ACM technical symposium on computer science education Atlanta, Georgia, pp. 145–150

Ashenhurst RL, Graham S (1987) ACM turing award lectures-the First Twenty Years. ACM Press, New York, NY

Carey T, Shepherd M (1988) Towards empirical studies of programming in new paradigms. Proceedings of the ACM 16th Annual Conference on Computer Science. (Atlanta, Georgia, United States), CSC '88. ACM Press, New York, pp. 72–78

Corder C (1990) Teaching hard teaching soft: a structured approach to planning and running effective training courses. Gower, Brookfield, VT

Denning PJ (2005) Is computer science science? Commun ACM 48(4):27–31

Denning PJ, Comer DE, Gries D, Mulder MC, Tucker A, Turner AJ, Young PR (1989) Computing as a discipline. Commun ACM 32(I):9–23

Dijkstra EW (1986) On a cultural gap. Math Intell 8(1):48–52

Floyd RW (1979) The paradigms of programming. Commun ACM 22(8):445–460

Haberman B, Ragonis N (2010) So different though so similar?-Or vice versa? Exploration of logic programming and of object oriented programming. Issues Inf Sci Inf Tech 7:393–402

Hazzan O (2008) Reflections on teaching abstraction and other soft ideas. Inroads- SIGCSE Bull 40(2):40–43

Hazzan O, Lapidot T (2004) Construction of a professional perception in the methods of teaching computer science course. Inroads- SIGCSE Bull 36(2):57–61

Hazzan O, Lapidot T (2006) Social issues of Computer Science in the Methods of Teaching Computer Science in the High School course. Inroads- SIGCSE Bull 38(2):72–75

Hazzan O, Har-shai G (2013) Teaching computer science soft skills as soft concepts. SIGCSE 2013—the 44th ACM technical symposium on computer science education. Denver, CO:59–64.

Hazzan O, Har-shai G (2014). Teaching and learning computer science soft skills using soft skills: the students' perspective. Proceedings of SIGCSE 2014—the 45th ACM technical symposium on computer science education. Atlanta, USA, pp. 567–572

Kuhn TS (1962) The structure of scientific revolution. University of Chicago, Chicago, IL

Lapidot T, Hazzan O (2003) Methods of teaching computer science course for prospective teachers. Inroads- SIGCSE Bull 35(4):29–34

Lave J, Wenger E (1991) Situated learning: legitimate peripheral participation (Learning in Doing: SOCIAL, Cognitive and Computational Perspectives). Cambridge University Press, Cambridge, UK

Microsoft Research (2005). Towards 2020 science. http://research.microsoft.com/en-us/um/cambridge/projects/towards2020science/downloads/T2020S_ReportA4.pdf. Accessed August 2014

Microsoft Research (2006) Towards 2020 science. Retrieved March 16, 2007, http://research.microsoft.com/towards2020science/background_overview.htm. Accessed July 14 2010

Ragonis N (2009) Computing pre-university: secondary level computing curricula. In: Wah B (ed) Wiley encyclopedia of computer science and engineering, (pp 632–648), 5(1), Wiley, Hoboken.

Sethi R (1996) Programming languages concepts & constructs, 2nd edn. Addison-Wesley, Reading, MA

Stolin Y, Hazzan O (2007) Students' understanding of computer science soft ideas: the case of programming paradigm. Inroads- SIGCSE Bull 39(2):65–69

Sukhoo A, Barnard A, Eloff MM, Van der Poll JA (2005) Accommodating soft skills in software project management. Issues Inf Sci Inf Tech 2:691–704

Tomayko J, Hazzan O (2004) Human aspects of software engineering. Charles River Media, Newton Center, MA

Tucker A, Noonan R (2002) Programming languages-principles and paradigms. McGraw Hill, New York, NY

Turkle S (1984) The second self: computer and human spirits. Simon and Shuster, New York, NY

Van Roy P, Haridi S (2004) Concepts, techniques, and models of computer programming/MIT Press, Cambridge, MA

Van Roy P, Armstrong J, Flatt M, Magnusson B (2003) The role of language paradigms in teaching programming. Proceedings of the 34th technical symposium on computer science education, Reno, Nevada, pp 269–270

Watt DA (1990) Programming language concepts and paradigms. Prentice Hall, Upper Saddle River, NJ

Wing J (2006) Computational thinking. Commun ACM 49(3):33–35

Research in Computer Science Education 4

4

> **Abstract**
>
> Computer science education research refers to different aspects such: students' difficulties, misconceptions, and cognitive abilities, to vary activities that can be integrated in the learning process, to the advantages of using visualization and animations tools, to the computer science teacher's role, and more. This meaningful sheered knowledge of the CS education community can enrich the prospective computer science teachers' perspective. The chapter exposes the MTCS students' to that reach resource, and practice ways in which they can be used in their future work. This knowledge may enhance lesson preparation, kind of activities developed for learners, awareness to learners' difficulties, ways to improve concept understanding, as well as testing and grading learners' projects and tests. We first explain the importance of exposing the students to the knowledge gained by the computer science education research community. Then, we demonstrate different issues addressed in such research works and suggest activities to facilitate with respect to this topic.

4.1 Introduction

Acquaintance and awareness of ongoing research in computer science education are important factors in the professional development of both prospective and in-service teachers. The Methods of Teaching Computer Science (MTCS) course provides a suitable opportunity to deliver to the prospective computer science teachers the message that their familiarity with research in computer science education may improve their work from different perspectives, e.g., enrich their learning and teaching processes, and broaden their understanding of their future pupils' prospective and their own understanding of computer science as well. Such familiarity will foster their professional development and allow them to make better pedagogical

© Springer-Verlag London Limited 2014   55
O. Hazzan et al., *Guide to Teaching Computer Science*,
DOI 10.1007/978-1-4471-6630-6_4

decisions. When reading and discussing computer science education research works in the MTCS course, focus should be placed both on the research itself (i.e., the research field, subject, questions, methods, and results) as well as on the research conclusions and recommendations. The discussions should highlight the fact that the prospective teachers can use these conclusions in their future work. In the continuation of the chapter, we first present a brief background of computer science education research, together with its applications in computer science learning and teaching processes. Second, we present several activities to be facilitated in the MTCS course, which introduce to the prospective computer science teachers the research in computer science education as well as its relevance for computer science teaching. In Chap. 6, we further expand the discussion on one central research topic in computer science education, i.e., learners' conceptions of computer science concepts, which deserve special attention due to its importance for computer science teaching and learning processes in general and in the high school in particular.

4.2 Research in Computer Science Education: What is it and Why and How is it Useful?

This section delves into the details of research on computer science education, addressing the following topics: categories of computer science education research, computer science education research on learning and teaching processes, and resources for computer science education research.

4.2.1 Computer Science Education Research Categories

Research in computer science education includes a variety of topics, which reflect a wide spectrum of interest. The focus of these topics has been changed over the years due to changes introduced in the discipline, the curriculum, programming languages, programming paradigms, computerized teaching tools, etc. Ragonis and Hazzan (2015, in review) explore the content and changes that took place in the last decade within the ITiCSE conference—one of the central venues for computer science education research. In what follows, we present several topics included in computer science education research together with illustrative research works associated with these topics:

- *Novice knowledge.* For example, Garner et al. (2005) investigate problems encountered by novice programmers; de Raadt (2007) reviews several recent studies which explored difficulties encountered by novices while learning programming and problem solving; Watson et al. (2014) have researched over the past 50 years predictors of programming performance. They found that a student's programming behavior is one of the strongest indicators of their performance, and suggest that research should be continued in that direction.
- *Concept understanding, e.g., variables, recursion, inheritance.* For example, Samurçay (1989) relates to cognitive difficulties which novice programmers

confront while learning the concept of variable; Kaczmarczyk et al. (2010) investigate learners' misconceptions of core basic computer science topics, e.g., memory models and assignment; Chaffin et al. (2009) suggest using a novel game that provides computer science learners the opportunity to write code, and base on visualization-based interaction to learn recursion by depth-first search of a binary tree; Denier and Sahraoui (2009) suggest visualizing inheritance in object-oriented programs to support learners' comprehension of this concept; Paul and Vahrenhold (2013) present results of the assessment of first-year students' misconceptions related to algorithms and data structures. They related to active and passive knowledge with respect to the teaching instruments used; Karpierz and Wolfman (2014) triangulate evidence for five misconceptions concerning binary search trees and hash tables, and design and validate multiple-choice concept inventory questions to measure the prevalence of these misconceptions.

- *Learning skills, e.g., problem solving, debugging, abstraction.* For example, Armoni (2009) analyzes computer science learners' abilities to reduce solutions of algorithmic problems; de Raadt et al. (2004) suggest a framework for instruction and assessment of problem-solving strategies; Edwards (2003) presents a vision for computer science education driven by the use of test-driven development; Murphy et al. (2008) present a qualitative analysis of debugging strategies of novice Java programmers; McCauleya et al. (2008) review the literature related to the learning and teaching of debugging computer programs; Miller et al. (2014) investigated the integration of computational thinking and creative thinking in CS1 to improve student learning performance according to Epstein's Generativity Theory.
- *Learning and teaching programming paradigms, e.g., functional, logical, procedural, object oriented.* For example, Van Roy et al. (2003) discuss the role of programming paradigms in teaching programming; Stolin and Hazzan (2007) investigate students' understanding of programming paradigm; Ragonis (2010) suggests a pedagogical approach for discussing fundamental object-oriented programming principles by using the ADT SET; Haberman and Ragonis (2010) explore teaching implication with respect to logic programming and object-oriented Programming; Bunde et al. (2014) illustrated a variety of relevant language paradigms by presenting parallel implementations of the Game of Life.
- Learning and teaching programming languages within a particular paradigm, e.g., with respect to object oriented programming languages: Smalltalk, Java, C#, Python. For example, Fleck (2007) discusses Prolog as the first programming language; Moritz and Blank (2005) explore a design-first curriculum for teaching Java in a CS1 course; Miller (2007) explores Python as a learning and teaching language.
- *Different teaching methods, e.g., laboratory work, projects-based learning, patterns.* For example, Hanks (2008) explores the advantages of pair programming while learning computer science; Soh et al. (2005) report on their framework for designing, implementing, and maintaining closed laboratories in CS1; Hauer and Daniels (2008) address open-ended group projects from a learning theory perspective; Forišek and Steinová (2010) suggest a set of didactic games and activities that can be used to illustrate and teach information theory concepts;

Simon (2013) used a peer instruction method that actively engages students in constructing their own learning, and found a main effect on the final exam grade.

- *Pedagogical usages of different computerized tools*, e.g., *Alice, BlueJ, Jeliot, Karel, Scratch.* For example, Kölling et al. (2003) elaborate on BlueJ and its pedagogy; Ben-Bassat et al. (2003) describe the usage of Jeliot animation environments to help novices understand basic concepts of algorithms and programming; Rodger et al. (2010) present pedagogical materials that expose students to computing by using Alice to develop projects, stories, games, and quizzes; Resnick et al. (2009) discuss the Scratch environment that enables simple creation of animations, games, and interactive art. Cross et al. (2014) describe the experiences of implementing dynamic program visualizations using the new viewer canvas in jGRASP by faculty and students in Java-based CS1 and CS2 courses.

- *Social issues in computer science teaching*, e.g., *diversity, ethics, and soft skills.* For example, Voyles et al. (2007) explore teachers' responses to gender differences and teachers' affect on achieving gender balance in their computer science classes; Dark and Winstead (2005) address potential changes in learners' conception when teaching ethics in computing-related fields; Baloian et al. (2002) describe common aspects and differences in the process of modeling the real world for applications involving tests and evaluations of cognitive tasks; Hazzan and Har-Shai (2013, 2014) presented a course aimed at advancing computer science students' soft skills by actively practicing those skills and gradually enableing students to construct their mental perception of these computer science soft skills.

- *Computer science teachers.* For example, Blum and Cortina (2007) describe a summer workshop for high school computer science teachers in which compelling material, which teachers can use in their classes to emphasize computational thinking, is provided; Ni (2009) explores the factors that influence computer science teachers' adoption of a new computer science curriculum; Mittermeir et al. (2010) report on a project that introduces both pupils and teachers to some principles of informatics, and, specifically, shows teachers that the concepts of informatics are not too difficult to teach; Brandes et al. (2010) describe a course for leading computer science teachers, and the results of the course and of the regional pedagogical workshops that these teachers facilitated for their peers; Lapidot and Ragonis (2013) describe a project to support high school computer science teachers in writing academic papers; Tashakkori et al. (2014) describe a unique 3-year project established to build teachers' research experience.

4.2.2   Computer Science Education Research on Learning and Teaching Processes

In this section, we focus on computer science education research on learning and teaching processes, which is especially relevant for high school computer science teaching.

4.2.2.1   Computer Science Education Research from the Learner's Perspective

Computer science education research works, that focus on learners, address computer science learners at different learning levels, e.g., junior high school, high school, and undergraduate level, and aim to deepen the understanding of how they conceive computer science concepts as well as their learning processes.

The inclusion of this research topic in the MTCS course aims to increase the prospective computer science teachers' awareness to learners' difficulties, to help them match teaching methods and tools for different learners' needs, and to guide them in designing the teaching process of different computer science concepts in a way that supports learners' learning processes (see also Chap. 11).

Two different aspects of learning processes seem to be interchangeable: mistakes and misconceptions (see also Chap. 6). A learner's answer, as well as learner's knowledge and understanding, can be either correct or incorrect. Though a mistake indicates, in most cases, lack of understanding, it should not necessarily be conceived negatively. Rather, an alternative point of view suggests that mistakes actually provide learners with the opportunity to correct their knowledge and improve their current understanding of the said concept, and therefore, further explanations and practicing is needed. From this perspective, the mere occurrences of learners' mistakes should encourage teachers to use different learning materials, methods, or tools in order to improve different cognitive skills. At the same time, however, a correct answer does not necessarily reflect understanding (see, e.g., Erlwanger 1973). Therefore, teachers should use a variety of learning assignments in order to reveal learners' misconceptions.

From the early research works in mathematics education and computer science education (e.g., Erlwanger 1973; Perkins and Martin 1986; Smith et al. 1993) we learn that in most cases, misconceptions are consistent and systematic, thought they could appear in multiple contexts; misconceptions are stable against attempts to change them; and misconceptions can sometimes block learning processes. Therefore, we should *not* use the term "mistake correction," which indicates that if we just correct the mistake we prevent the misconception and improve the learners' understanding. Rather, we should look for misconceptions and help learners correct them by looking for and addressing their source. This alternative approach leads to an improved concept understanding as well as correct answers in the future. In other words, teaching processes which are sensitive to misconceptions support the learning process since they guide learners in constructing their knowledge by themselves and in accordance with their own knowledge structures (see also the constructivist approach in Chap. 2). Clearly, this perspective should be delivered and discussed with the prospective computer science teachers in the MTCS course. It will be further elaborated in Chap. 6, where learners' alternative conceptions are explored in depth.

In addition to the rich aspects mentioned above, we mention a technological aspect in general and the integration of computerized tools in particular. Even the tendency to agree that, in some cases, different visualization and animation environments extenuating learners' understanding, in some cases it misleading (Ragonis and Ben-Ari 2005) and adds difficulty (Krauskopf 2012). Teachers have to be aware of all those learning obstacles as well.

4.2.2.2   Computer Science Education Research from the Teacher's Perspective

There are many teaching situations in which a professional teacher should think and act as a researcher. In this spirit, one of the targets of the MTCS course is to educate the prospective computer science teachers to become attentive to their pupils, to reveal their pupils' conceptions, and to check and track pupils' understandings during the entire learning process, and not just at its end by a final exam. In this spirit, one of the messages delivered in the MTCS course is that the "teacher as researcher" perspective plays a significant role in becoming a professional computer science teacher. From a broader perspective, research works that examine topics related to computer science teachers intend to deepen the knowledge and understanding of the computer science teachers' work. This research addresses teachers' disciplinary knowledge, teachers' pedagogical knowledge, teachers' pedagogical-content knowledge (PCK), teachers' coping with curricular changes, and teachers' use of different teaching tools.

Computer science education research may contribute to computer science teachers' knowledge and professional development in general and to the prospective computer science teachers in the MTCS course in particular, in several ways, as is outlined in what follows.

- *Becoming a member of the computer science education community.* The familiarity with computer science education research is one component of becoming a member of the computer science education community. This component includes, among other things, acquaintance with the collective wisdom gained by the researchers of the community and with its common and accepted terminology.
- *Increases teachers' awareness of learning processes.* Teachers' familiarity with research in computer science education increases their awareness to different learners' conceptions, difficulties, and experiences with computer science concepts. This type of knowledge belongs to one of the central categories of Shulman's (1986) teacher knowledge base model—i.e., knowledge of learners— which can assist teachers, for example, in the preparation process of different teaching units (see, e.g., Chap. 11). Further, by exposing teachers to the research in computer science education, teachers can learn from other teachers' experiences, and may be encouraged by the fact that learners in other classes exhibit similar difficulties to those manifested by their pupils, and that they cope with similar challenges to those faced by other teachers.
- *Construction teachers' pedagogical-content knowledge (PCK).* Learning from computer science education research is valuable also for teachers' construction of their PCK, another central category of Shulman's teacher knowledge base model (Shulman 1986). This knowledge includes answers to questions such as: What examples, analogies, and demonstrations are suitable for the explanation

of specific computer science topics? What is an appropriate way to organize the teaching sequence of a specific topic? What is the source for specific learners' difficulties? Which learning/teaching strategies are shown (by research findings) to work better in a specific teaching situation? A significant layer—T—was added to the PCK: TPCK, which means technological pedagogical-content knowledge. TPCK is a framework to understand and describe the kinds of knowledge teachers need for an effective pedagogical practice in a technology enhanced learning environment. This relates to the uses of any environments and specifically with respect to teaching computer science (e.g., Ben-Bassat Levy and Ben-Ari, 2007, 2008; Lee 2011; Ioannou and Angeli 2013; Mouza et al. 2014). In addition to finding answers to these questions in the research literature on computer science education, computer science teachers can find in these literature recommendations for diagnosis exercises, teaching methods, teaching tools, etc.

- *Adopting a researcher's point of view.* Teachers' exposure to research in computer science education expands their acquaintance with various research tools. This acquaintance may, in turn, enrich their pedagogical toolbox and let them be more attentive to specific characteristics of their pupils. Specifically, such knowledge may enable them to be sensitive to their pupils' needs, conceptions, difficulties, and cognitive skills, and accordingly, to enrich their pedagogical toolbox and improve their pedagogical skills. Here are two examples for the implementation of these ideas: (1) Lapidot and Ragonis (2013) describe a project conducted with leading high school computer science teachers to support their preparation and submission process of a paper or a poster to an academic conference; (2) Tashakkori et al. (2014) established a 3-year project to build research experience for teachers. The objective was to provide in-service high school teachers and community college faculty to work with faculty mentors and their graduate and undergraduate assistants to conduct research in data analysis and mining, visualization, and image processing. It can be seen that both projects were built based on the belief that exposing teachers to the computer science education research and actively writing a research paper has the potential to advance their teaching skills in many ways.

4.2.3   Resources for Computer Science Education Research

Many resources exist for computer science education research. It is important to expose the prospective computer science teachers to these resources, and maybe, even further, to explore with them a potential contribution to one of these resources. We list here several such resources:

- The Association for Computing Machinery (ACM) digital library[1]
- The Special Interest Group on Computer Science Education (SIGCSE) conference proceedings[2]. The papers are published also in the ACM digital library

---

[1] See http://portal.acm.org/dl.cfm

[2] The 2014 conference website is: http://sigcse2014.sigcse.org/

- The Innovation and Technology in Computer Science (ITiCSE) conference proceedings[3]. The papers are published also in the ACM digital library
- The Informatics in Secondary Schools: Evolution and Perspective (ISSEP) conference proceedings[4]
- Inroads—ACM SIGCSE Bulletin[5]
- The ACM Inroads magazine, launched in 2010[6]
- The Computer Science Education Journal (CSEJ)[7]
- The ACM Transactions on Computing Education (TOCE)[8]
- The Workshop in Primary and Secondary Computing Education (WiPSCE) website[9]

4.3   MTCS Course Activities

Activities 25–29, to be facilitated by the prospective computer science teachers in the MTCS course, explore the research in computer science education from a pedagogical perspective. The instructor of the MTCS course can decide what activities fit his or her class as well as on the order of the activities.

Activity 25: Exploration of a Computer Science Education Research Work on Learners' Understanding of Basic Computer Science Topics
The general target of this activity is to present to the students an example of a research work that deals with learners' understanding of variables—one of the basic computer science concepts, and specifically, to focus on alternative conceptions, mistakes, and misconceptions related to variables.

The details of this activity are presented in Activity 44 (Sect. 6.3), which explores learners' alternative conceptions. When this activity is facilitated in the MTCS course, it is recommended to highlight the direct contribution of research in computer science education to teachers' knowledge on learners' conceptions and the consecutive actions that they can take based on this understanding.

---

[3] The 2014 conference website is: http://iticse2014.it.uu.se/

[4] The 2014 conference website is: http://www.issep2014.org/

[5] See http://portal.acm.org/toc.cfm?id=J688

[6] See http://portal.acm.org/browse_dl.cfm?linked=1&part=magazine&idx=J1268&coll=portal&dl=ACM&CFID=77578246&CFTOKEN=38064848

[7] See http://www.tandf.co.uk/journals/titles/08993408.asp

[8] See http://portal.acm.org/browse_dl.cfm?linked=1&part=transaction&idx=J1193&coll=portal&dl=ACM

[9] See http://www.wipsce.org/2014/index.php, the 2014 conference website

Activity 26: The Computer Science Education Research World

The purpose of the activity is to practice with the prospective computer science teachers the reading of research papers in computer science education. It is suitable especially for prospective computer science teachers who have not read research papers before and are not familiar with computer science education research. The activity is based on three stages.

- Stage A: Intuitive thinking on computer science education research, class discussion

The students are asked to answer the following questions presented to them by the instructor. Their answers are written on the board.

- In your opinion, what do researchers in computer science education do?
- In your opinion, what questions/issues/topics are researchers in computer science education interested in?

- Stage B: Planning a research in computer science education, group work

The students are asked to choose one of the questions/issues/topics listed on the board and to suggest a research plan for its exploration. They can be guided by suggesting them to relate to questions such as: What would a researcher do in order to find answers for the said research questions? Which research tools can be used? What would a researcher do in order to improve his or her understanding of the said topic?

- Stage C: Class discussion

Each group presents its suggestion. If time allows, it is recommended to discuss these suggestion and illuminate core research aspects. In this discussion, however, it is important to remember that (in most cases) the students do not have a previous research experience in computer science education.

Activity 27: Looking into a Research Work on Novices' Difficulties, Homework

The target of this homework activity is to expose the prospective computer science teachers to different aspects of novice computer science learners' mental processes, as they arise from the research papers published in the field. For this purpose, in this activity, the students read for the first time a research paper in computer science education.

The topic of novice learners is chosen for this activity because it is important that teachers become familiar with novice difficulties and understanding and since it is reasonable to assume that the prospective computer science teachers will teach such learners in the future. Further, we should remember that, in fact, each learner is a novice learner when learning a new topic; from this perspective, the prospective computer science teachers should be aware to the fact that even a trivial assignment (for experts) may raise substantial difficulties for novices and therefore, an in-depth investigation of learners' conceptions should not be neglected.

Specifically, in this homework, each student is required to read one research paper from a given list of papers on novice computer science learners, and to submit a report based on a given list of guiding questions (see Table 4.1 below). It is recommended that the instructor verifies that all the papers in the list are selected. It is also recommended to publish the students' reports, for example, in the course website.

Comments:

- The list of papers presented in the worksheet is just an example. Each instructor can construct a new list either on novice learners or any other topic discussed by the research community on computer science education.
- The homework includes two optional tasks.
  - The first optional task aims to encourage the students to look for additional research paper recourses in order to first, increase their familiarity with these research resources and second, to practice how to find a paper that relates to a specific topic.
  - The second optional task (mini-research) should be facilitated only with prospective computer science teachers who have some background in educational research.

Table 4.1 Homework on research in computer science education

Homework on computer science education research
The following list presents research papers on difficulties encountered by computer science novice learners[a]
Fluery (1993)
Joni and Soloway (1986)
Pea (1986)
Perkins and Martin (1986)
SamurÇay (1985)
Spohrer and Soloway (1986)
Choose one paper and address the following tasks:
List the three main points that the paper stresses
List three ideas that came into your mind while reading the paper. You can relate to what you found most interesting, what increased your curiosity, or issues you would like to read more about
Optional: Find in the literature on computer science education another research paper that is somehow connected to the paper you chose. Indicate the paper title, authors, and abstract and explain in what sense the two papers are related
Mini-research (optional): Choose one of the papers and reconstruct the research described in the paper

[a] The list of papers was collected by Haberman, Levy and Lapidot as part of a survey on novice programmers' difficulties, published in Hebetim—Journal of the Israeli National Center for Computer Science Teachers, June 2002, pp. 8–38 (In Hebrew)

Activity 28 which deals with learners' misconceptions naturally fits for the MTCS course. A similar activity, however, can be facilitated in any computer science class since it is expected that any computer scientist be familiar with common mistakes. In both cases, the computer science subject on which the activity is based can be changed according to the context in which the activity is facilitated.

Further, the analysis of correct and incorrect solutions can promote the understanding of any computer science concept. Such an analysis can be carried out in two stages:

a. Classification of given (both correct and incorrect) solutions for a given problem
b. Discussion of the correct and the incorrect solutions: with respect to correct solutions, to find their uniqueness; with respect to incorrect solutions, to discuss what is wrong and to correct them

---

Activity 28: The Teacher as a Researcher

The target of this activity is to give the prospective computer science teachers an opportunity to identify learners' misconceptions. This activity, which gives the prospective teachers a simple context to investigate and examine the source of learners' mistakes, can be viewed as a first experience in adapting a researcher's point of view. Such an experience may enable them in the future to help their students repair their own conceptions and improve their understanding.

The students receive a set of solutions for a given problem that addresses a basic use of one-dimensional arrays. First, they are asked to solve the question by themselves. Second, they should identify correct and incorrect solutions from a given collection of solutions. Third, for each of the incorrect solutions they should suggest a possible source for the incorrect answer, i.e., to speculate what the learner who presented the said solution does not understand or (even better) does not understand well. Later, they are asked to suggest additional examples for incorrect solutions for problems that result from misconceptions they assume learners may have. Finally, several reflective activities take place.

Overall, the activity includes six parts and is based on individual work, work in pairs and class discussion. As can be observed, it addresses several pedagogical aspects and topics, such as, alternative conceptions (see Chap. 6), reflection (see Chap. 5), constructivism (see Chap. 2), and different types of questions (see Chap. 9).

• Stage A: Solving a problem, individual work

Each student is given a worksheet with the problem (see Table 4.2) and is asked to solve it and to keep all the drafts. In addition, the students are asked to present a list of computer science concepts that in their opinion are manifested in this question.

Table 4.2 Worksheet with an introductory problem

A problem
Write an algorithm that returns *true* if all array A's members are equal and *false* if they are not. Array A is of size N (the array indexes are 1-N)
Solve the problem
Keep the drafts you wrote during the process of solution development
When you are finished, put your solutions aside and ask for the next task

Since a solution development to a given problem enables problem solvers capture the problem domain more meaningfully and differently than checking a solution (see Stage B), the students are asked to solve the problem by themselves. It is recommended to collect their solutions for the later stages of the activity (C and E).

• Stage B: Evaluating different solutions, individual work

Usually, teachers design questions to check their pupils' overall understanding; at the same time, researchers' questions are more refined, as they intend to identify learners' conceptions, including unexpected ones. This stage intends to let the students adopt the researcher's point of view. Accordingly, in this stage of the activity, the students first experience a task that a computer science teacher performs on a daily basis—determine whether a solution is correct or incorrect, and second, practice the researcher's work—looking for the source of the mistake, i.e., identifying learners' misconception(s)/alternative conception(s).

The students are given six solutions to the above problem and are asked first to classify them into correct and incorrect solutions and second, for each incorrect solution, to speculate what might be the source of the mistake; i.e., with respect to the (wrong) solutions, to hypothesize what might be the pupils' misconception(s). See Table 4.3.

• Stage C: Discussion on Stage B answers, work in pairs

In this stage, the students work in pairs, reflect on their previous work, compare their solutions, discuss and argue, and try to convince each other. It is hoped that during this experience, they construct a partial mental image of what researchers in computer science education do.

Specifically, the students are asked to discuss and reflect in pairs on their classifications of the answers presented in Stage B (Table 4.3). In addition, they are asked to exchange their own solutions from Stage A and check each other's solution. The specific instructions are presented below:

1. Discuss your conclusions from the previous stage. Compare your answers and, if needed, elaborate/change/correct them.
2. Exchange with your partner your own solution to the problem. Check your partner's solution.
 – If it is correct: How can you develop your future students' capabilities in problem-solving processes?
 – If it is incorrect: Is it similar to one of the incorrect solutions presented in the worksheet you worked on (Stage B/Table 4.3)?

- Stage D: The meaning of learners' mistakes, class discussion

After the students finish working in pairs, it is recommended to facilitate a class discussion to help them assimilate the process they went through. The discussion can address the following issues:

- Mistakes that have been discovered by the students, focusing on possible misconceptions that could lead to these mistakes.
- Differences between tasks that ask to "find an incorrect solution" and tasks that ask to "explore why a mistake occurred."
- The importance of this kind of investigation for (prospective) computer science teachers. For example, such an investigation can help teachers in lesson preparation and can direct them to change the teaching order of different topics, to use different teaching materials, and to use animation or other demonstration tools.

- Stage E: Taking the researcher's perspective, work in pairs

This stage is optional and can be facilitated by the instructors of the MTCS course who choose to deepen the discussion on research tools.

The prospective computer science teachers are asked to use the drafts of their own solution developed in Stage A, and by adopting a researcher's perspective, to discuss in pairs what, in their opinion, can be learnt from this examination. For example, the collection of intermediate drafts of a solution can help a researcher in investigating learners' mental processes, examining different directions a learner chose, guessing what caused a learner to change directions during the development process, etc.

- Stage F: Reflection, individual work

The students are asked to reflect on the activity. The reflection can be either an open and spontaneous reflection or a guided reflection as presented in Table 4.4. The selection of the kind of the reflection partially depends on students' previous awareness to reflection (see Sect. 5.6). Additional reflection questions can address the kind of activity itself, e.g., how were you contributed from the discussion in pairs? What was the most difficult stage in this activity and why, etc.

Activity Summary:

As mentioned before, the target of this activity is to expose the prospective computer science teachers to a new perspective—the researcher's point of view—by offering them to consider different approaches when learners' mistakes are examined. This exposure is expected to broaden their considerations as computer science teachers, mainly when they design teaching materials. Specifically, they are expected to take the learners' perspective, to consider what they understand, what might mislead their understanding, and which learning tools can help novice learners acquire new knowledge.

In addition to the pedagogical contribution of this activity, it also contributes to the prospective computer science teachers' PCK since they are exposed to different mistakes computer science learners may encounter in the context of arrays and logical conditions.

Table 4.3 Worksheet on uncovering alternative conceptions

Worksheet—uncovering alternative conceptions
High school students were asked to solve the same problem you just solved Write an algorithm that returns true if all array A items are identical and false if they are not identical
1. In what follows, six students' solutions are presented. For each solution
Determine whether it is correct or incorrect
If it is incorrect
Describe the mistake
Hypothesize the source of the mistake and write it down
Suggest how you would help a student understand the mistake
2. Present additional incorrect solutions based on a misconception you assume that novice learners may have

Solution 1	Solution 2
are-equals (A, N)	are-equals (A, N)
ok ← true	ok ← true
for i from 1 to N do	for i from 1 to N-1 do
if (A[i]≠A [i+1]) then	if (A[i]≠A[i+1]) then
ok ← false	ok ← false
return ok	Else
	ok ← true
	return ok
Solution 3	Solution 4
are-equals (A, N)	are-equals (A, N)
ok ← true	ok ← true
for i from 1 to N-1 by 2 do	for i from 1 to N-1 by 2 do
if (A[i]≠A[i+1]) then	if (A[i]≠A[i+1]) then
ok ← false	ok ← false
return ok	for i from 2 to N-1 by 2 do
	if (A[i]≠A[i+1]) then
	ok ← false
	return ok
Solution 5	Solution 6
are-equals (A, N)	are-equals (A, N)
ok ← true	count ← 0
for i from 1 to N-1 do	for i from 2 to n do
if (A[i]≠A[i+1]) then	if (A[i]=A[i+1]) then
ok ← false	count ← count+1
return ok	if (count=N) then
	return true
	Else
	return false

Table 4.4 Reflective task on the researcher's point of view

Reflection on the "teacher as a researcher" activity
What did you learn from the "teacher as a resellarcher" activity?
How were you contributed from this activity with respect to your future work as a computer science teacher?
In your opinion, can you use any part of the activity with your future learners?

Activity 29: Reflection on Reading a Computer Science Education Paper, Homework

The target of this homework activity is to expose the prospective computer science teachers to a comprehensive paper that deals with the uniqueness of computer science teaching (Gal-Ezer and Harel 1998). It is suggested that the work on this paper may develop the professional identity of the prospective computer science teachers. See Table 4.5.

The paper explores different types of knowledge that a computer science teacher should master, and is actually related to the essence of the PCK with respect to computer science teaching (Shulman 1986, 1990). Since the paper relates to a bird's eye view of the field, it is recommended to give this activity to the students as a summary task of the subject of research in computer science education (i.e., after at least one of the previous activities has been facilitated). It is also recommended to follow the students' individual work with a class discussion, including the instructor's contribution.

Table 4.5 Reading and reflection on a fundamental computer science education paper

Worksheet
Read the paper: Gal-Ezer and Harel (1998)
Choose two of the subjects explored in the paper that in your opinion are the most significant for a computer science teacher. Explain your selection
In your opinion, does the "teacher as a researcher" activity fit any of the categories that the paper proposes? If it does, indicate the specific category and explain your choice. If in your opinion the "teacher as a researcher" activity does not belong to any of the categories mentioned in the paper, suggest a new category to which it belongs, explain its uniqueness, and define it Here is a suggestion for a possible extension of the task: Read the paper: Gal-Ezer and Zur (2013) Summarize its main messages

The assimilation of the advantages of being aware and attentive to the computer science education research developments and trends should not be restricted to the kind of activities presented in this chapter. A meaningful adaptation will be gained if the use of computer science education papers is integrated along the entire course. In fact, any development of teaching materials or activity can start with posing the question "what can we learn from other educators?" An example can be found in Sect. 11.3.1 "Planning the Teaching of a Study Unit About One-Dimensional Array."

References

Armoni M (2009) Reduction in computer science: a (mostly) quantitative analysis of reductive solutions to algorithmic problems. JERIC 8(4):1–30

Baloian N, Luther W, Sánchez J (2002) Modeling educational software for people with disabilities: theory and practice. Proceedings of the 5th International ACM Conference on Assistive Technologies, pp 111–118

Ben-Bassat Levy R, Ben-Ari M (2007). We work so hard and they don't use it: acceptance of software tools by teachers. In Proceedings of the 12th annual SIGCSE conference on Innovation and technology in computer science education (ITiCSE '07). ACM, New York, USA, pp 246–250

Ben-Bassat Levy R, Ben-Ari M (2008). Perceived behavior control and its influence on the adoption of software tools. SIGCSE Bull 40(3):169–173

Ben-Bassat Levy R, Ben-Ari M, Uronen PA (2003) The jeliot 2000 program animation system. Comput Educ 40(1):1–15

Blum L, Cortina TJ (2007) CS4HS: An outreach program for high school CS teachers. ACM SIGCSE Bull 39(1):19–23

Brandes O, Vilner T, Zur E (2010) Software design course for leading CS in-service teachers. Proceedings of ISSEP, Lecture Notes in Computer Science, vol 5941, 49–60

Bunde DP, Graf M, Han D, Mache J (2014). Parallel programming paradigms illustrated (abstract only). In Proceedings of the 45th ACM technical symposium on Computer science education (SIGCSE '14). ACM, New York, USA, pp 722–722

Chaffin A, Doran K, Hicks S et al (2009) Experimental evaluation of teaching recursion in a video game. Proceedings of the 2009 ACM SIGGRAPH Symposium on Video, pp 79–86

Cross J, Hendrix D, Barowski L, Umphress D (2014) Dynamic program visualizations: an experience report. In Proceedings of the 45th ACM technical symposium on Computer science education (SIGCSE '14). ACM, New York, USA, pp 609–614

Dark MJ, Winstead J (2005) Using educational theory and moral psychology to inform the teaching of ethics in computing. Proceedings InfoSecCD, pp 27–31

Denier S, Sahraoui H (2009) Understanding the use of inheritance with visual patterns. Proceedings of the 3rd International Symposium on Empirical Software Engineering and Measurement, pp 79–88

de Raadt M (2007) A review of Australasian investigations into problem solving and the novice programmer. Comput Sci Educ 17(3):201–213

de Raadt M, Toleman M, Watson R (2004) Training strategic problem solvers. ACM SIGCSE Bull 36(2):48–51

Edwards SH (2003) Rethinking computer science education from a test-first perspective. 18th Annual ACM SIGPLAN OOPSLA Conference, pp 148–155

Erlwanger SH (1973) Benny's conception of rules and answers in IPI mathematics. JCMB 1(2):7–26

Fleck A (2007) Prolog as the first programming language. ACM SIGCSE Bull 39(4):61–64

Fluery AN (1993). Student beliefs about Pascal programming. *J Educ Comput Res* 9(3): 355–371

Forišek M, Steinová M (2010) Didactic games for teaching information theory. In: Vahrenhold J (ed) Lecture notes computer science, vol 5941. Springer, Berlin, pp 86–99

Gal-Ezer J, Harel D (1998) What (else) should CS educators know?. Communic ACM 41(9):77–84

Gal-Ezer J, Zur E (2013) What (else) should CS Educators Know?—Revisited, WiPSCE '13, Aarhus, Denmark, pp 84–87

Garner S, Haden P, Robins A (2005) My program is correct but it doesn't run: A preliminary investigation of novice programmers' problems. Proceedings of the 7th Australasian Conference on Computing Education, vol 42, pp 173–180 (Young A, Tolhurst D (eds))

Haberman B, Ragonis N (2010) So different though so similar?—Or vice versa? Exploration of the logic programming and the object-oriented programming paradigms. Iss Informing Sci Inf Technol 7:393–402

Hanks B (2008) Problems encountered by novice pair programmers. JERIC 7(4):1–13

Hauer A, Daniels M (2008) A learning theory perspective on running open ended group projects (OEGPs). Proceedings 10th Conference on Australasian Computing Education, vol 78, pp 85–91 (Australian Compu. Soc., Darlinghurst, Australia)

Hazzan O, Har-Shai G (2013) Teaching computer science soft skills as soft concepts. In Proceeding of the 44th ACM technical symposium on Computer science education (SIGCSE '13). ACM, New York, USA, pp 59–64

Hazzan O, Har-Shai G (2014) Teaching and learning computer science soft skills using soft skills: the students' perspective. In Proceedings of the 45th ACM technical symposium on Computer science education (SIGCSE '14). ACM, New York, USA, pp 567–572

Joni, S.A., Soloway, E. (1986). But my program runs! Discourse rules for novice programmers. J Educ Comput Res 2(1):95–128

Kaczmarczyk LC, Petrick ER, East JP et al (2010) Identifying student misconceptions of programming. Proceedings 41st ACM Technical Symposium on Computer Science Education. pp 107–111

Karpierz K, Wolfman SA (2014) Misconceptions and concept inventory questions for binary search trees and hash tables. In Proceedings of the 45th ACM technical symposium on Computer science education (SIGCSE '14). ACM, New York, USA, pp 109–114

Kölling M, Quig B, Patterson A et al (2003) The BlueJ system and its pedagogy. Comput Sci Educ 13(4):249–268

Krauskopf K, Zahn C, Hesse FW (2012) Leveraging the affordances of youtube: the role of pedagogical knowledge and mental models of technology functions for lesson planning with technology. Comput Educ 58(4):1194–1206

Lapidot T, Ragonis N (2013) Supporting high school computer science teachers in writing academic papers. In Proceedings of the 18th ACM conference on Innovation and technology in computer science education (ITiCSE '13). ACM, New York, USA, pp 325–325

Lee YL (2011) The development of technological pedagogical content knowledge for science learning with a three-dimensional interactive computer simulation. Ph. D. Dissertation, University of Washington, Seattle, WA, USA Advisors Mark Windschitl AAI3472171

Ioannou I, Angeli C (2013) Teaching computer science in secondary education: a technological pedagogical content knowledge perspective. In Proceedings of the 8th Workshop in Primary and Secondary Computing Education (WiPSE '13). ACM, New York, USA, pp 1–7

McCauleya R, Fitzgeraldb S, Lewandowskic G et al (2008) Debugging: a review of the literature from an educational perspective. Comput Sci Educ 18(2):67–92

Miller B (2007) Exploring python as a learning and teaching language. J Comput Small Coll 22(3):262–263

Miller D, Soh LK, Chiriacescu V, Ingraham E, Shell DF, Hazley MP (2014) Integrating computational and creative thinking to improve learning and performance in CS1. In Proceedings of the 45th ACM technical symposium on Computer science education (SIGCSE '14). ACM, New York, USA, pp 475–480

Mittermeir RT, Bischof E, Hodnigg K (2010) Showing core-concepts of informatics to kids and their teachers. In: Vahrenhold J (ed) Lecture notes in Computer Science, vol 5941. Springer, Berlin, pp 143–154

Moritz SH, Blank GD (2005) A design-first curriculum for teaching Java in a CS1 course. ACM SIGCSE Bull 37(2):89–93

Mouza C, Karchmer-Klein R, Nandakumar R, Ozden SY, Hu L (2014) Investigating the impact of an integrated approach to the development of preservice teachers' technological pedagogical content knowledge (TPACK). Comput Educ 71:206–221

Murphy L, Lewandowski G, McCauley R et al (2008) Debugging: the good, the bad, and the quirky: a qualitative analysis of novices' strategies. Proceedings 39th SIGCSE Technical Symposium Computer Science Education, pp 163–167

Ni L (2009) What makes CS teachers change? Factors influencing CS teachers' adoption of curriculum innovations. Proceedings 40th ACM Technical Symposium Computer Science Education, pp 544–548

Paul O, Vahrenhold J (2013) Hunting high and low: instruments to detect misconceptions related to algorithms and data structures. In Proceeding of the 44th ACM technical symposium on Computer science education (SIGCSE '13). ACM, New York, USA, pp 29–34

Pea, R.D. (1986). Language-independent conceptual "bugs" in novice programming. J Educ Comput Res 2(1):25–36

Perkins DN, Martin F (1986) Fragile knowledge and neglected strategies in novice programmers. In: Soloway E, Iyengar S (eds) Empirical studies of programmers. Ablex Pub, Norwood, pp 213–229

Ragonis N (2010) A pedagogical approach to discussing fundamental object-oriented programming principles using the ADT SET. ACM Inroads 1(2):42–52

Ragonis N, Ben-Ari M (2005) On understanding the statics and dynamics of object-oriented programs. SIGCSE Bull 37(1):226–230

Ragonis N, Hazzan O (2015, in review) What Are Computer Science Educators Interested In? The CASE of SIGCSE Conferences. Submitted to SIGCSE

Resnick M, Maloney J, Monroy-Hernández A et al (2009) Scratch: programming for all. Commun ACM 52(11):60–67

Rodger SH, Bashford M, Dyck L et al (2010) Enhancing K-12 education with Alice programming adventures. Proceedings on Innovation and Technology in Computer Science Education, pp 234–238

Samurçay, R. (1985). Learning programming: an analysis of looping strategies used by beginning students. For Learn Math 5(1):37–43

Samurçay R (1989) The concept of variable in programming: its meaning and use in problem-solving by novice programmers. In: Soloway E, Spohrer JC (eds) Studying the novice programmer. Lawrence Erlbaum Associates, New Jersey, pp 161–178

Shulman LS (1986) Those who understand: knowledge growth in teaching. J Educ Teach 15(2):4–14

Shulman LS (1990) Reconnecting foundations to the substance of teacher education. Teach Coll Record 91(3):300–310

Simon B, Parris J, Spacco J (2013) How we teach impacts student learning: peer instruction vs. lecture in CS0. In Proceeding of the 44th ACM technical symposium on Computer science education (SIGCSE '13)

Smith JP III, diSessa AA, Roschelle J (1993) Misconceptions reconceived: a constructivist analysis of knowledge in transition. J Learn Sci 3(2):115–163

Soh L, Samal A, Nugent G (2005) A framework for CS1 closed laboratories. JERIC 5(4):2

Spohrer, J.C., Soloway, E. (1986). Analyzing the high frequency bugs in novice programs. In Soloway E, Iyengar S (eds). Empirical studies of programmers. Ablex Pub.: Norwood, pp 230–251

Stolin Y, Hazzan O (2007) Students' understanding of computer science soft ideas: the case of programming paradigm. ACM SIGCSE Bull 39(2):65–69

Tashakkori RM, Parry RM, Benoit A, Cooper RA, Jenkins JL, Westveer NT (2014) Research experience for teachers: data analysis & mining, visualization, and image processing. In Proceedings of the 45th ACM technical symposium on Computer science education (SIGCSE '14). ACM, New York, USA, pp 193–198

Van Roy PA, Flatt M et al (2003) The role of language paradigms in teaching programming. Proceedings of the 34th SIGCSE Technical Symposium Computer Science Education, pp 269–270

Voyles MM, Haller SM, Fossum TV (2007) Teacher responses to student gender differences. Proceedings of the 12th Annual SIGCSE Conference on Innovation & Technology in Computer Science Education, pp 226–230

Watson C, Li FWB, Godwin JL (2014). No tests required: comparing traditional and dynamic predictors of programming success. In Proceedings of the 45th ACM technical symposium on Computer science education (SIGCSE '14)

Problem-Solving Strategies

5

Abstract

Problem solving is one of the central activities performed by computer scientists as well as computer science learners. However, computer science learners often face difficulties in problem analysis and solution construction. Therefore, it is important that computer science educators are aware of these difficulties and acquire appropriate pedagogical tools to help their learners gain experience in these skills. This chapter is dedicated to such pedagogical tools. It presents several problem-solving strategies to address in the MTCS course together with appropriate activities to mediate them to the prospective computer science teachers.

5.1 Introduction

Since programming is a problem-solving process, problem-solving skills must be a core idea of any introductory computer science course. However, whereas the teaching of programming languages is usually well-structured within a curriculum, the development of learners' problem-solving skills is largely implicit and less structured. Therefore, we find it relevant to include this topic in this guide.

Problem solving is a complex mental process. This inspection can be easily observed by looking, for example, at the problem-solving techniques listed at Wikipedia: abstraction, analogy, brainstorming, divide and conquer, hypothesis testing, lateral thinking, means-ends analysis, method of focal objects, morphological analysis, reduction, research, root cause analysis, and trial-and-error.

Problem-solving processes are common to many disciplines. In the mathematics education research community, for example, an intensive discussion takes place about problem-solving processes and techniques, as well as on learners' difficulties and ways of teaching problem-solving strategies (see, e.g., Polya 1957; Schoenfeld 1983).

© Springer-Verlag London Limited 2014
O. Hazzan et al., *Guide to Teaching Computer Science*,
DOI 10.1007/978-1-4471-6630-6_5

In some cases, learners develop problem-solving strategies by themselves. For example, children invent simple addition and subtraction strategies long before they learn arithmetic at school. But, without some formal instruction of effective strategies, even the most inventive learner may resort to unproductive trial-and-error problem-solving processes. Hence, it is important to teach problem-solving strategies and to guide teachers how to teach their pupils this cognitive tool. Due to the centrality of problem-solving processes in the discipline of computer science, this assertion is especially relevant for computer science education.

The attention for the need to explicitly teach problem-solving strategies is expressed also in the development of computerized tools designed for that purpose. For example, to meet the learners' individual problem-solving strategies, their solutions, and their perceptions, a research project done by Kiesmüller (2009) draws design guidelines for a learning environment that can identify and categorize several problem-solving strategies automatically. The target is to give the learner feedback and guidelines according to his or her own problem-solving skills and to guide them accordingly to their thinking processes tendency. Hasni and Lodhi (2011) suggest guidelines to design lab exercises that guide learners through the problem-solving process. The emphasis is on defining detailed steps that scaffold the learners throughout the process. These steps are reflected in Sects. 5.3, 5.4, 5.5 and can be interpreted as an implementation of this approach.

This chapter highlights the perspective according to which computer science educators should be aware of problem-solving strategies and acquire pedagogical tools to help their learners acquire this kind of skills. The chapter is dedicated to these pedagogical tools and presents several agreed-upon illustrative problem-solving strategies to address in the MTCS course, and mediate them by appropriate activities. It also addresses some new ideas related to looking at computer science problem solving.

Our discussion focuses mainly on algorithmic problems[1], since most of the prospective computer science teachers will probably teach these kind of problems in their high school teaching.

With respect to most of the ideas presented in this guide, it is important to deliver to the students enrolled in the MTCS course that all the strategies explored in the context of this chapter can be implemented in any programming paradigm and/or programming language. Further, it is important to highlight that problem-solving strategies are not restricted to programming tasks.

5.2 Problem-Solving Processes

A basic problem-solving process, in any discipline, starts with outlining the problem requirements, and terminates with outlining a solution that, in some cases, is expressed by a sequence of steps (an algorithm) that solves the problem. In com-

---

[1] An algorithmic problem is defined by what is given—the initial conditions of the problem, and its goals—the desired state, what should be accomplished. An algorithm problem can be solved with a series of actions formulated formally either by pseudo code or a programming language.

puter science, in many cases, the algorithm is coded into a programming language and is tested by the code execution. The difficult stages, however, lie in between: how to move from the requirement understanding to the problem solution. These intermediate stages can be viewed as a discovery process; therefore, problem-solving processes are sometimes treated as a creative process, as an art.

Commonly recognized stages of problem-solving processes are listed below. Needless to say that in most cases, problem-solving processes are not linear and intertwine the different stages described below:

1. *Problem analysis.* Understand and identify the problem, understand what the problem is about. If one does not understand the problem, one cannot proceed and solve it (see Sect. 5.3).
2. *Alternative consideration.* Think about alternative ways to solve the problem.
3. *Choosing an approach.* Choose an appropriate approach to solve the problem.
4. *Problem decomposition.* Decompose the problem into subtasks.
5. *Algorithm development.* Develop the algorithm in stages according to the recognized subtasks (see Sect. 5.4).
6. *Algorithm correctness.* Check the algorithm correctness (see Sect. 5.5).
7. *Algorithm efficiency.* Calculate the algorithm efficiency.
8. *Reflection.* Reflect on and analyze the process you went through, conclude what can be improved in future problem-solving processes (see Sect. 5.6).

Though defined rules how to perform these stages do not exist, we can scaffold computer science learners by several methods/strategies that may help them cope with this process. From this perspective, problem-solving processes can be classified as soft ideas (see Chap. 3). Activity 30 addresses several problem-solving techniques in computer science.

Activity 30: Problem-Solving Techniques in Computer Science
The target of this activity is to increase students' familiarity with different problem-solving techniques within the computer science discipline. This activity can be the first one in the discussion about problem solving, but can also be facilitated as a summary activity for this subject.[2]

Students are asked to work in small groups (of 2–4 students) and to look at a list of problem-solving techniques given either in Wikipedia (as mentioned at the beginning of Sect. 5.1) or in other resources (e.g., Vasconcelos 2007).

Each group selects five problem-solving techniques which are commonly used in computer science problem-solving processes, and for each one of them describes at least two computer science class situations in which it can be addressed.

---

[2] In advanced computer science classes, it is relevant to mention that in computer science, in addition to the development of problem-solving strategies, special emphasis is placed also on non-solvable problems (see Chap. 9).

For example, the *Analogy* technique can be addressed when a new problem is introduced. In such cases, it is worthwhile to examine first similar pre-solved problems, and based on the analogy between the two (or more) problems, to derive the solution for the new problem.

The discussion facilitated in the MTCS course based on this activity can relate to the fitness of different problem-solving techniques to the discipline of computer science. It is also worth discussing connections between different problem-solving techniques in computer science, e.g., connections between stepwise refinement (Sect. 5.4.2) and algorithmic patterns (Sect. 5.4.3).

5.3   Problem Understanding

Problem understanding is the first stage of problem-solving processes that leads to the identification of the problem characteristics. It can be based on analogies to similar problems and sometimes it yields some generalizations.

In the case of algorithmic problems, this stage can start by recognizing the input categories of the problem and the selection of the required output category for each input category, respectively. The identification of the input categories actually indicates that the problem is understood. Each input category is usually treated in a different subtask, or at least addressed when the algorithm correctness is checked. Extreme cases should be analyzed as well: sometimes they are included in a specific category; in other cases, they are treated separately. Activity 31 focuses on this aspect of problem-solving processes.

Activity 31: Examination of Representative Inputs and Outputs
The target of this activity is (a) to increase students' awareness of a problem-analysis process guided by the examination of representative inputs and outputs and (b) to acquire pedagogical tools to deliver the importance of this stage to their future pupils.

- Stage A: Problem development, work in pairs
  The students are asked to develop two problems that illustrate the importance of examining representative inputs and outputs. One problem should be about basic conditions and loops; it should be relatively simple and fit learners at their early stages of computer science learning. The second problem should address more advanced computer science topics, such as, a two-dimensional array, a list, a stack, or a queue. It is important to emphasize that even in advanced stages of computer science learning, and maybe even especially in these stages, it is important to allocate the needed time for this first stage of solving algorithmic problems.
- Stage B: Presentations and discussion

Each pair presents to the course plenum one of the problems it developed. For each problem, the pair should explain why it is important to start its solving process with the examination of inputs and outputs.

If possible, the course instructor should choose the problems that the pairs present in a way that enables discussion on different aspects and topics.

It is recommended to publish all the developed questions in the course web site for future usages either in the MTCS course (see Activity 32 below, for example) or in the students' future high school teaching.

5.4 Solution Design

Some of the major difficulties novice learners face, when engaged in problem-solving processes, are concerned with the early stages of the solution design (Soloway 1986; Reed 1999; Robins et al. 2003). In what follows we present three strategies that may be employed in this stage of the solution design, i.e., defining the problem variables, stepwise refinements and algorithmic patterns, together with relevant activities that examine these strategies from a pedagogical perspective. Needless to say that the order by which the strategies are presented here is not necessarily the order by which they should be taught. Alternatively, the order by which they are taught and used, as well as the level of depth on which each of them is discussed, depend on the learning stage and the kind of problem in hand. For example, in simple cases, it is reasonable to start with choosing the needed variables; in more complex cases, it is worth starting developing the problem solution by stepwise refinements.

5.4.1 Defining the Problem Variables

The examination of a given problem's inputs and their corresponding outputs, clarifies the problem to the problem solver. The next stage is to define the variables needed in order to solve the problem. Activities 32–34 examine solution design from a pedagogical perspective.

Activity 32: Choosing the Problem Variables
This activity aims to deliver the message that a deep consideration of the variables needed for solving a given problem is a crucial stage in the problem-solving process since it directs the solution implementation.

- Stage A: Problem analysis, work in pairs
  Students are asked to further analyze the problems they developed in Activity 31 and to set the variables needed to solve them. It is recom-

mended that each pair will work both on the problems it developed and on the problems developed by at least one more pair. In this way, each student examines four problems. The problem exchange between the pairs also contributes to the discussion that takes place in Stage B.

- Stage B: Discussion between pairs
  Each two pairs that exchanged their problems discuss together each of the four problems. In this discussion, they are asked to:
  a. Compare their solutions
  b. List guidelines for a teacher who evaluates learners' investigation of the variables needed for solving a given problem
- Stage C: Presentations and discussion in the course plenum
  If time allows, a discussion can be facilitated in the course plenum in which the teacher guidelines, formulated by all groups, are examined.

Activities 33–34 focus on roles of variables, which is a recently introduced approach that can be utilized in teaching programming to novice learners.

The role of a variable is defined according to the dynamic character of the variable, embodied by the succession of values that variable gets, and how the new values assigned to the variable relate to other variables. For example, in the role of *a stepper*, a variable is assigned a succession of values known in advance as soon as the succession starts. The role concept, however, is not concerned with how a variable is used in the program; a stepper is a stepper whether it is used to index elements in an array or to count the number of input values (Ben-Ari and Sajaniemi 2003; Sajaniemi 2005).

The concept of roles of variables is concerned with the deep structure of the program. The name of the variable, its place within an expression, relations between expressions and assignments, and the control statements, are important factors in role assignment; the surface structure of the program, primarily its syntactic structure, is less relevant to the role concept.

Activities 33 and 34 focus on the roles of variables. They can be facilitated in the MTCS course, either in the class or as homework assignments.

Activity 33: Roles of Variables—Discovery Learning and Reflection
The students are directed to the Web site of roles of variables,[3] to learn the relevant concepts and practice them by themselves. Then, the students are asked to reflect on their experience.

---

[3] The roles of variables home page (http://www.cs.joensuu.fi/~saja/var_roles/ very rich and contains different kinds of educational resources).

- Stage A: Learning, work in pairs
  1. Learn the "roles of variables" concept by reading the "Introduction to the Roles of Variables."[4]
  2. Perform the activity "Try to Classify Variables Yourself."[5]
  3. Try to use "The Role Advisor."[6]
- Stage B: Reflection on Stage A, individual work

  At this stage, the students are directed to reflect on what they learnt from two perspectives: as learners of a new concept and as future computer science teachers (see Chap. 2). Since it is a reflection activity (see Sect. 5.6), it is recommended to carry it out individually. Specifically, the students' tasks are:
  1. What new kinds of knowledge did you gain from this exploration?
  2. Do you agree with the interpretation of the "roles of variables" concept? Do you agree with the categorization of the variable roles?
  3. Did you enjoy this kind of learning? Did you enjoy doing the "check yourself" activity?
  4. Address advantages and disadvantages of this kind of learning.
  5. As a future high school computer science teacher, reflect on using this kind of discovery learning in high school classes: Do you think that this interpretation for variable roles should be integrated to scaffold learners' problem-solving skills? Do you think that pupils can learn it the same way as you did?

Activity 34: Roles of Variables—Examination of the Roles of Variables Through the Research Lens

The target of this activity is to expose the students to current computer science educational research on the roles of variables (see also Chap. 4). They look at relevant research developments and investigate resources that can serve them in their future teaching in the high school. This work may also enhance their curiosity as future researchers.

The students' tasks are:

1. Explore the following resources in the Roles of Variables home page:
   - Why Roles[7]

---

[4] See http://www.cs.joensuu.fi/~saja/var_roles/role_intro.html.

[5] See http://www.cs.joensuu.fi/~saja/var_roles/try.html.

[6] See http://cs.joensuu.fi/~pgerdt/RAE/.

[7] See http://www.cs.joensuu.fi/~saja/var_roles/why_roles.html.

- Using Roles of Variables in Programming Education[8]
- Literature on the Roles of Variables[9]
2. What are your impressions with respect to the research work that examines the roles of variables concept?
3. Write 5–10 ideas that, while reading, you found interesting, curious, and innovative. Explain your choice of each of these ideas.

5.4.2   Stepwise Refinement

The main purpose of the stepwise refinement design methodology is to first obtain an overview of the structure of the problem and of the relationships among its parts, and then to address specific and complex issues related to the implementations of the various subparts (Wirth 1971; Dijkstra 1976). Stepwise refinement is a top-down methodology since it progresses from the general to the specific. An alternative approach is the bottom-up methodology that progresses from the specific to the general. These two approaches can be considered as complementing each other. In both cases, the problem is divided into sub-problems; the main difference lies in the direction of the mental process that guides the solution construction. Here, we focus on the top-down approach.

Solutions produced by stepwise refinements possess a natural modular structure, which (a) is easier to develop and to check, (b) increases the solution readability, and (c) enables to use the solution of sub-problems of the full solution for solving other problems as well.

The stepwise refinement approach can be applied in any software development process. Throughout this process, an initial representation of some solution, on a high level of abstraction, is gradually refined through a sequence of intermediate representations that eventually yield a final program in some programming language (Batory et al. 2004). While the initial representation employs notations and abstractions that are appropriate for the addressed problem, the development proceeds in a sequence of small steps that each of them refines some aspect of the representation produced in previous steps.

The approach is usually associated with Niklaus Wirth, who formulated the main principles of stepwise refinements:

- In each step, one or several instructions of the given program are decomposed into more detailed instructions; a step can involve a simultaneous refinement of both data structures and operations.
- Every refinement step implies some design decisions. It is important that these decisions be explicit, and that the programmer be aware of the underlying criteria and of the existence of alternative solutions.

---

[8] See http://www.cs.joensuu.fi/~saja/var_roles/teaching.html.

[9] See http://www.cs.joensuu.fi/~saja/var_roles/literature.html.

The importance of the stepwise refinements process increases as the problem complexity elevates. But, as it turns out, it is also useful for solving basic problems, such as finding the maximum of three numbers, as is discussed, for example, in Reynolds et al. (1992).

Activities 35 and 36 focus on stepwise refinement.

Activity 35: Practicing Stepwise Refinement—Break-Down Problem Solutions into Subtasks

The students continue working on the problems they developed in Activity 31, and are asked to break down the problem solutions into subtasks.

Activity 36: Practicing Stepwise Refinement—Analyze a List of Problems

The students receive a list of problems and are asked to analyze them comprehensively, considering inputs–outputs categories, variable selection, and the implementation of stepwise refinements.

Learning from an expert as an apprentice approach is very well appreciated in many professions (e.g., architecture, music). Based on the premise that in reality problem-solving skills are developed through experiencing and practicing over a period of time, Arshad (2009) presents a pedagogical approach to teach problem solving by using the thinking-aloud strategy. Specifically, students learn the skill of problem solving by closely observing and listening to an "experienced programmer". The evidence show that think-aloud problem solving is extremely effective for female students. These research findings probably correspond with the personal experience of each programmer. A quite frequent scenario is in which one receives a logical problem defined in terms of an algorithm or a program, and while explaining the problem-solving approach (either to a classmate or to an educator or an educator explains to one of his or her students), he or she finds the solution. Activity 37aims at highlighting the potential of the think-aloud problem-solving strategy and practicing it in a structured manner.

Activity 37: Practicing Think-Aloud Problem Solving

The target of this activity is to enrich students' considerations while solving a problem by listening to a peer who solves the same problem (specifically, with MTCS classmates, or, in the case of a programming-oriented course, with classmates or a two students from different academic years).

The pair gets a problem; one of them is the "expert" and the second is the "apprentice". The "expert" solves the problem and speaks out loudly all his or her consideration, while following all the problem-solving steps; the "apprentice" listens. It is important to stick to the scenario in which one speaks and the other listens. Only after the "expert" finishes laying out all his or her considerations, questions, thoughts, reflection, correction, thinking again, etc., the "apprentice" is permitted to response and ask questions. It is recommended to alternate the pair roles, and repeat the same activity. To summarize the activity, it is recommended to allocate time to a guided reflective process (see Sect. 5.6) either freely or by a list of pre-prepared questions. After the reflective session takes place, each pair presents its conclusions in the course plenum.

5.4.3   Algorithmic Patterns

Algorithmic patterns are entities that combine design elements and mathematical aspects (Ginat 2004). Within its context, a pattern denotes an expert solution to a recurring design or a programming problem (East et al. 1996; Soloway 1986; Wallingford 1996; Astrachan et al. 1998; Reed 1999; Proulx 2000; Muller et al. 2004). Similar to other patterns, algorithmic patterns represent examples of elegant and efficient solutions of recurring algorithmic problems. Thus, in fact, an algorithmic pattern is an abstract model of an algorithmic process that solves a specific problem and can be matched or modified and then, integrated into the solution of different problems.

Muller et al. (2007) argue that when facing an unfamiliar problem, learners often do not know how to start solving it, and experience difficulties in (a) recognizing similarities between problems and in transferring ideas from previously solved problems to new ones; and (b) observing the essence of a problem and in identifying its components and the relationships among them. Therefore, in such cases, they tend to reinvent the wheel and develop a solution from scratch. These difficulties, however, do not usually result from some misunderstanding, but rather may be a consequence of a poor organization of algorithmic knowledge.

The use of algorithmic patterns, which is based on the problem analysis, can assist learners in this process of solution development for algorithmic problems. This process starts with the recognition of the solution components and of similar problems whose solution is already known. Then, pattern(s) should be modified for the solution of the current problem and be integrated into the full solution. In most cases, there is a need to combine different patterns to develop the needed solution. In general, however, the target of dealing with patterns in the learning of computer science is to strengthen learners' problem-solving abilities which go beyond programming.

Ragonis (2012) suggests an activity aimed at imparting algorithmic patterns to prospective computer science teachers, and presents findings with respect to difficulties prospective teachers' experience when coping with patterns in general and composing recursion patterns in particular. These findings support the inclusion of the topic in computer science teacher preparation programs. Activities 38–40, which examine algorithmic patterns from several perspectives, are derived from that research. In addition, see Activity 101 in Sect. 12.8 about patterns of recursive list processing. It is recommended to ask the students to work on these activities in pairs.

Prior to the facilitation of these activities, instructors of the MTCS course should verify that the students are familiar with algorithmic patterns either by a self-learning process or by employing a variety of teaching methods as is presented in this Guide. Here are several resources that can be used for this purpose: Clancy and Linn (1999); Reed (1999); Muller (2005); Muller et al. (2007); Ginat (2009). The instructor of the MTCS should select from these resources a set of patterns to be used in Activity 38.

Activity 38: Practicing Algorithmic Patterns—Question Design for Given Patterns
A list of patterns is presented. Each pair of students chooses two different patterns and:

1. For each pattern
   - Develops an algorithm problem such that its solution uses this pattern.
   - Develops "a story problem" (see Chap. 9) such that its solution uses this pattern.
   - Develops a problem such that its solution requires a slight change of the pattern.
2. Develops two problems such that their solution requires the combination of the two patterns.

Activity 39: Practicing Algorithmic Patterns—Pattern Composition for Using Specific Abstract Data Types
The students are asked to develop three algorithmic patterns that use specific abstract data types, such as, a linked list, a queue, a stack, or a tree.

Activity 40: Practicing Algorithmic Patterns—Worksheet Design for Guiding Learners Using Patterns
The students are asked to develop a worksheet that guides novice computer science learners how to use patterns. The targets of the worksheet are (a) to introduce the idea of patterns; (b) to direct learners in what cases it is appropriate to consider using patterns; and (c) to guide learners to reflect on their use of patterns.

The students should also articulate the guidelines they followed in the design process of this worksheet.

5.5 Debugging

After a solution has been designed and constructed, its correctness should be examined. Similar to many computer science concepts, checking the correctness of
a solution has both theoretical and technical aspects. In high school classes, it is
sufficient, in most cases, to examine the solution correctness technically with representative inputs; in more advanced computer science teaching situations, this discussion should be expanded and includes also correctness theoretical aspects.

One of the aspects related to solution examination is debugging. It is accepted
that the debugging processes (either with a debugger or without it) may promote
learners' understanding of computer science (cf., for example, Lieberman 1997;
Spohrer and Soloway 1986). This assertion is based on the fact that while students
consider how their program should be tested and debugged if necessary, they actually reflect on how they implemented computer science concepts and rethink their
programming process and the decision they made during the program development
(Laakso et al. 2008). One may assume that debugging is a skill that comes with
common sense; nevertheless, it is known as a meaningful obstacle in developing
computer programs and even crucial in project development. In recent years, some
attention is given to the skill of debugging in general and to teaching debugging
in particular. For example, Nagvajara and Taskin (2007) advocate the importance
of teaching debugging skills throughout digital design courses, especially during
the introductory courses, and present teaching techniques for developing students'
debugging skills both for introductory and advanced digital design courses. These
techniques emphasize incremental design stages, test stimuli, observation techniques, and critical thinking. Ahrendt et al. (2009) describe a new course "Testing,
Debugging, and Verification" which connects formal approaches with real-world
development techniques in a non-traditional and novel way. Their approach views
formalization of specifications as the basis for debugging and test generation tools.
Vírseda et al. (2011) describe an innovative methodology, based on a logic teaching
tool, to prepare students to use logic as a formal proof technique, including declara-
tive debugging of imperative programs, which should be considered at the basis of
a good software development.

Here we focus on debugging due to its relevance to high school teaching in the
contexts of solution examination. It is also highly relevant, of course, in more advanced computer science teaching situations, as well as for software practitioners
during the development of software projects.

Unlike other areas in which errors are treated as a negative phenomenon, in CS,
errors and debugging are a fact of life and, accordingly, deserve special attention.

According to the constructivist theory, learning consists of a debugging process
of one's own knowledge. Specifically, learning (of any topic) starts by building
a preliminary (sometimes erroneous) model and then proceeds with shaping it,
through a successive refinement process, into another (more correct) model. In order to start the refinement process, one must acknowledge the fact that one's initial
mental model might be erroneous. With respect to software development, each program we write reflects the way in which we conceive the problem we face (Papert

1980); debugging is a means that enables us to come closer to the target. Awareness of this role of debugging can improve one's understanding of learning processes.

Like Papert (1980), we believe that learning to debug can *change students' negative attitudes toward errors* and increase their awareness with respect to the importance of mistakes in learning processes. This view is even strengthened among students who are accustomed to being evaluated by automated accepted grading systems. There, mistakes (or errors) lower their grade and the opportunity to learn from the mistakes and to improve the entire problem-solving process is neglected. Some researchers expanded this view point and suggest a teaching-learning process which is based on "learning from mistakes". Some of them are based on class discussions (Ginat 2003, 2008; Ginat and Shmalo 2013), and some integrate a computerized tool to scaffold the process of acquiring self-debugging skills (Seta et al. 2006; Raman et al. 2012). The main issue is that when errors become a legitimate topic for discussion, it becomes easier to address the connections between learning processes and misconceptions. That leads to our recommendation to explicitly teach debugging, since learning about debugging might improve students' *understanding of their own thinking processes*.

The inclusion of debugging in the MTCS course can deliver important messages to the prospective computer science teachers. Specifically, it has several advantages listed below:

- *Awareness to novice programmers' debugging strategies*: Future computer science teachers must be aware of the fact that debugging strategies are important from the early steps of novice programmers' work and that, accordingly, computer science teachers should deal with debugging-related issues in their daily work. Carver and McCoy (1988) indicate that "children typically respond ineffectively to debugging situations. Many of them panic and call the teacher insistently, making no attempts to correct the situation until help arrives. Others quickly erase their code and begin again without ever understanding their error." Consequently, computer science teachers should be aware of different ways by which to approach such debugging-related situations.
- *Computer science teachers must be good debuggers:* Computer science teachers must acquire good debugging skills in order to be able to understand students' buggy programs. Good debugging skills enable them to guide learners in their debugging processes. Although student-teachers are familiar with debugging strategies from their own computer science studies, the attention given to debugging in the MTCS course might further improve their debugging skills.
- *Recognition of the opportunity to lead discussions on mistakes*: Through the discussion on debugging, computer science teachers should realize the opportunity they have to change negative attitudes towards errors they (or their future pupils) may have. Discussions on mistakes may deepen the understanding of the taught concepts, and may include both content concepts (such as, loops or arrays), and soft skills content (such as, problem-solving strategies).
- *Automatic testing:* In current development processes in both academia and industry, it is more and more common to base the development process on auto-

matic testing which eases and improves debugging strategies. Some IDEs even support inherently such development processes. When appropriate, this aspect can be added to the discussion. Nevertheless, in all cases, the resistance sometimes exists to testing and debugging (see, e.g., Hazzan and Leron 2006) should not be neglected.

Activities 41 and 42 about debugging processes, to be facilitated in the MTCS course, highlight pedagogical merits of debugging processes (see also the musical debugging (Activity 65) in Chap. 8).

Activity 41: Examination of the Debugging Process
Students are asked to write a program that solves a specific problem and to document both their programming process and their debugging process. Such an experience not only increases learners' awareness of their debugging strategies, but also fosters their reflective skills.

Activity 42: Development of a Lab Activity about Debugging, Teamwork
The students are asked to work in small groups (of 2–3 students) and to develop a lab activity that aims to scaffold learners' debugging skills.

This kind of activity is challenging since it requires dealing with skills rather than with concepts. Since it is a very open task, it is reasonable to assume that a large, varied, and colorful collection of artifacts will be created. This collection can be used for discussions in the MTCS course as well as for the students' future work as high school computer science teachers.

We recommend educators to deliver the following messages throughout the debugging activities: (a) debugging is an excellent example of connections exist between content and pedagogy; (b) debugging is a fact of life, both in learning and programming processes, and therefore is inevitable; (c) mistakes are not a negative phenomenon, as long as one knows how to learn from them and how to fix them; (d) tools and strategies that support debugging exist and knowing how to use them can improve one's debugging skills; and (e) teachers should address debugging-related issues and teach their students to value the importance of debugging processes in software development processes.

5.6   **Reflection**

Reflective thinking refers to rethinking and analysis methods of previous mental processes or actual behaviors. Reflection is an important tool in learning processes in general and in high-order cognitive processes, such as problem-solving processes, in particular. This assertion is based on the recognition that reflection provides learners with an opportunity to step back and think about their own thinking and by doing so to improve their problem-solving skills. Reflective thinking is a learned process that requires time and ongoing practice (see also Chaps. 2 and 13).

During problem-solving processes, reflection can take place at different times:

- *Before* starting solving the problem: After reading the problem, while planning the solving approach, it is worth reflecting on similar previously solved problems in order to identify relevant algorithmic approaches, patterns, etc.
- *While* solving the problem: During the solution development, reflection refers to inspection, control, and supervision. For example, when a difficulty arises or when a mistake is identified, it is worth reflecting on their sources. Schön (1983) calls this process reflection *in*-action.
- *After* solving the problem: When the solution is completed, reflection assesses and examines the process performance. Such reflection enables to draw conclusions from the problem-solving process, and to learn about the strategic decisions made during its implementation. Schön (1983) calls this process reflection *on*-action.

In what follows we present a list of representative questions that can guide before-and-after reflective processes; some of them relate to cognitive aspects and other— to affective aspects. Even though the questions are organized according to the types of reflection, most of them can serve (with slight changes) in each of the three reflective stages.

Questions before starting the problem-solving process

- How can I estimate the question difficulty? Is the question difficult/easy? Why do I think so?
- Do I face any difficulty in understanding any part of the problem? What part is unclear?
- Did I previously solve similar problems? What are these problems? What are the similarities?

Questions after completing the problem-solving process

- Is the solution complete?
- Why did I choose this direction to solve the problem? Did I make reasonable decisions?
- What should I change in future similar situations?
- Was solving this problem easy or difficult for me? Why?
- Could I solve the problem differently? How?

The integration of reflective processes in teaching processes is a creative task. Activity 43 illustrates how reflective activities can become a learning process both from the learners' and the teachers' perspective. Specifically, teachers can reflect on the teaching process of critical concepts that were difficult for learners to grasp and on his or her usage of different teaching tools. In addition, an ongoing reflection may increase teachers' awareness to pupils' perspective.

Activity 43: Reflective Activity in Computer Science Education
The activity is based on the following case study that should be presented first to the students.

> After a computer science high school class had written a test on relatively advanced computer science topics (like linked lists or pushdown automaton), its teacher realized that the pupils' achievements in the test were low and that their solutions did not indicate the expected understanding of these concepts.

The following four stages are based on this case study. It is recommended to summarize each stage with a reflective discussion and to publish students' products in the course Web site.

- Stage A: Teacher's reflection on test failure
  The students are divided into small groups and are asked to:
  1. Write down a list of reasons that can explain the pupils' relatively low achievements.
  2. Classify the list of reasons into two groups: Teacher-oriented reasons and learner-oriented reasons. Reflect on the classification process. Was the classification process evident? In what cases did you hesitate? Why did you hesitate?
  3. Suggest at least five questions that can guide a teacher's reflective process on his or her class' failure.
- Stage B: Design of a reflective activity for a class after a test failure
  The students are divided into small groups and are asked to suggest how they, as computer science teachers, can use reflective processes to support their pupils' learning with respect to two aspects: (1) pupils' learning processes and (2) pupils' understanding of the learned concepts.
  Specifically, the students' task is to design a reflective activity for a high school class that aims at leading the pupils to reflect on their own strengthens and weaknesses, while taking into the consideration the two above-mentioned aspects.
- Stage C: Reflection on the reflective activity designed by a teacher after the test failure. The continuation of the case study is presented to the students: The teacher of the said class decided to use a reflective process to scaffold pupils' learning processes, to improve their understanding of the said computer science concepts, and to improve the test achievements. For this purpose, the teacher asked the pupils to answer the following questions:

What problems did you solve while learning toward the test? How did you solve them? What were your difficulties while writing the test? Did you face these difficulties only during the test or did you face them also while learning toward the test? If you faced these difficulties while learning toward the test, what did prevent you to deepen your understanding? If you faced the difficulties only during the test, try to speculate why you did not face them before.

The students are asked to work in small groups and to:

1. Classify the reflective questions into two groups: questions that relate to pupils' learning processes and questions that relate to pupils' understanding of the learned concepts.
2. Design at least two additional reflective questions for each class of reflective questions.
3. In your opinion, how can such reflective questions contribute to learners' future learning processes in general and problem-solving processes in particular?

- Stage D: Analysis of the entire reflective activity designed by the teacher after the test failure

This stage is based on the analysis of the next episode of the case study.

After the pupils had written their own reflection based on the previous reflective questions (see Stage C), they were asked to further accomplish the next two phases:

1. To design by themselves a test on the same contents of the test they took with the same structure. The teacher rationale for this task was that the development of meaningful questions requires deep understanding of the subject matter. To accomplish this task, the pupils were also asked to solve the questions they composed.
2. To reflect on their current knowledge by considering the following leading questions provided by the teacher: Did you overcome your previous difficulties? How did you overcome these difficulties? Do you feel ready to take a repeated test? With respect to what concepts you still feel unconfident? What do you think about the computer science concepts you learned—interest vs. boring; important vs. unnecessary; difficult vs. simple?

The students' task is to analyze the advantages and disadvantages of this kind of teaching–learning process.

For example, we mention the following advantages: pupils are active, take responsibility on their learning processes and understanding, think and focus on what concepts they did not understand, take the teacher's perspective, and reflect on different affective issues, such as their interests, priorities, concerns, and confidence. Nevertheless, it should be remembered that such an activity sets several pedagogical challenges, such as it demands pupils' cooperation, requires time resources, and it requires creativity which may be difficult for some learners.

Beyond the examination of this activity from the pupils' perspective, it is important to discuss this activity also from the teacher's perspective. In addition, it is important to highlight the fact that this kind of activity promotes nontraditional interactions between the teacher and the pupils and has the potential to promote the class-learning climate and future teaching–learning processes.

References

Ahrendt W, Bubel R, Hahnle R (2009) Integrated and tool-supported teaching of testing, debugging, and verification. In: Nuno Oliveira (eds) Proceedings of the 2nd International Conference on Teaching Formal Methods (TFM '09), Jeremy Gibbons and Jos\&\#233. Springer-Verlag, Berlin, pp 125–143

Arshad N (2009) Teaching programming and problem solving to CS2 students using think-alouds. SIGCSE Bulletin 41, 1(March 2009), pp 372–376

Astrachan O, Berry G, Cox L, Mitchener G (1998) Design patterns: an essential component of CS curricula. Proceedings of SIGCSE, pp 153–160

Batory D, Sarvela JN, Rauschmayer A (2004) Scaling stepwise refinement. IEEE Trans Softw Eng 30(6):355–371

Ben-Ari M, Sajaniemi J (2003) Roles of variables from the perspective of computer science educators. University of Joensuu, Department of Computer Science., Technical. Report, Series A-2003-6

Carver S, McCoy (1988) Learning and transfer of debugging skills: applying task analysis to curriculum design and assessment. In: Mayer RE (ed) Teaching and learning computer programming, multiple research perspectives. Lawrence Erlbaum Associates, Inc., Chapter 11

Clancy MJ, Linn MC (1999) Patterns and pedagogy. Proc SIGCSE' 99:37–42

Dijkstra EW (1976) A discipline of programming. Prentice-Hall, New Jersey

East JP, Thomas SR, Wallingford E, Beck W, Drake J (1996) Pattern-based programming instruction. Proceedings of ASEE Annual Conference and Exposition, Washington, DC

Ginat D (2003) The greedy trap and learning from mistakes. SIGCSE Bull 35(1):11–15

Ginat D (2004) Algorithmic patterns and the case of the sliding delta. SIGCSE Bull 36(2):29–33

Ginat D (2008) Learning from wrong and creative algorithm design. SIGCSE Bull 40(1):26–30

Ginat D (2009) Interleaved pattern composition and scaffolded learning. Proceedins of the 14th annual ACM SIGCSE conference on Innovation and Technology. in Computer Science Education—ITiCSE '09, Paris, France, pp 109–113

Ginat D, Shmalo R (2013) Constructive use of errors in teaching CS1. In Proceedings of the 44th ACM technical symposium on Computer science education (SIGCSE '13). ACM, New York, NY, USA, pp 353–358

Hasni TF, Lodhi F (2011) Teaching problem solving effectively. ACM Inroads 2, 3(August 2011), pp 58–62

Hazzan O, Leron U (2006) Why do we resist testing? Syst Des Front—Exclus Front Cover Syst Des 3(7):20–23

Kiesmüller U (2009) Diagnosing learners' problem-solving strategies using learning environments with algorithmic problems in secondary education. Trans. Comput. Educ. 9, 3, Article 17 (September 2009), 26 pages

Laakso MJ, Malmi L, Korhonen A, Rajala T, Kaila E, Salakoski T (2008) Using roles of variables to enhance novice's debugging work. Iss Inf Sci Inf Technol 5:281–295

Lieberman H (1997) The debugging scandal and what to do about it (special section). Comm ACM 40(4):27–29

Muller O (2005) Pattern oriented instruction and the enhancement of analogical reasoning. Proceedings of the First International Workshop on Computing Education Research ICER '05, Seattle, WA, USA, pp 57–67

Muller O, Haberman B, Averbuch H (2004) (An almost) pedagogical pattern for pattern-based problem solving instruction. Proceedings of the 9th Annual SIGCSE Conference on Innovation Technology in Computer Science Education, pp 102–106

Muller O, Ginat D, Haberman B (2007) Pattern-oriented instruction and its influence on problem decomposition and solution construction. ACM SIGCSE Bull 39(3):151–155

Nagvajara P, Taskin B (2007) Design-for-debug: a Vital Aspect in Education. In Proceedings of the 2007 IEEE International Conference on Microelectronic Systems Education(MSE '07). IEEE Computer Society, Washington, DC, USA, pp 65–66

Papert S (1980) Mindstorms: children, computers and powerful ideas. Basic Books Inc., New York

Polya G (1957) How to solve it. Doubleday and Co., Inc., Garden City

Proulx VK (2000) Programming patterns and design patterns in the introductory computer science course. Proceedings of SIGCSE, pp 80–84

Ragonis N (2012) Integrating the teaching of algorithmic patterns into computer science teacher preparation programs. In Proceedings of the 17th ACM annual conference on Innovation and technology in computer science education (ITiCSE '12). ACM, New York, NY, USA, pp 339–344

Raman K, Svore KM, Gilad-Bachrach R, Burges CJC (2012) Learning from mistakes: towards a correctable learning algorithm. In Proceedings of the 21st ACM int. conf. on Information and knowledge management (CIKM '12). ACM, New York, NY, USA, pp 1930–1934

Reed D (1999) Incorporating problem-solving patterns in CS1. J Comput Sci Edu 13(1):6–13

Reynolds RG, Maletic JI, Porvin SE (1992) Stepwise refinement and problem solving. IEEE Softw 9(5):79–88

Robins A, Rountree J, Rountree N (2003) Learning and teaching programming: areview and discussion. Comput Sci Edu 13(2):137–172

Sajaniemi J (2005) Roles of variables and learning to program. In: Jimoyiannis A (ed) Proceedings of the 3 rd Panhellenic Conference Didactics of Informatics. University of Peloponnese, Korinthos, Greece. http://cs.joensuu.fi/~saja/var_roles/abstracts/didinf05.pdf. Accessed 3 July 2010

Schoenfeld AH (1983) Episodes and executive decisions in mathematical problem-solving. In: Lesh R, Landaue M (eds) Acquisition of mathematics conceptsand processes. Academic Press Inc., New York

Schön DA (1983) The reflective practitioner. Basic Books, New York

Seta K, Kajino T, Umano M, Ikeda M (2006) An ontology based reflection support system to encourage learning from mistakes. In: Deved V (ed) Proceedings of the 24th IASTED international conference on Artificial intelligence and applications (AIA'06). ACTA Press, Anaheim, pp 142–149

Soloway E (1986) Learning to program = learning to construct mechanisms and explanations. CACM 29(1):850–858

Spohrer JG, Soloway E (1986) Analyzing the high frequency bugs in novice programs. In: Soloway E, Iyengar S (eds) Empirical studies of programmers. Ablex, Norwood, pp 230–251

Vasconcelos J (2007) Basic strategy for algorithmic problem solving. http://www.cs.jhu.edu/~jorgev/cs106/ProblemSolving.html. Accessed 2 June 2010

Vírseda RdV Orna EP Berbis E Guerrero SdL (2011) A logic teaching tool based on tableaux for verification and debugging of algorithms. In: Blackburn P, van Ditmarsch H, Soler-Toscano F, Manzano M (eds) Proceedings of the third international congress conference on tools for teaching logic (TICTTL'11). Springer-Verlag, Berlin, pp 239–248

Wallingford E (1996) Toward a first course based on object-oriented patterns. Proceedings SIGCSE, pp 27–31

Wirth N (1971) Program development by stepwise refinement. CACM 14(4):221–227. http://sunnyday.mit.edu/16.355/wirth-refinement.html. Accessed 13 Nov 2010

Learners' Alternative Conceptions

6

Abstract

This chapter focuses on learners' alternative conceptions. Since prospective teachers in general, and prospective computer science teachers in particular, face difficulties in gaining the notion of alternative conceptions, it is important to address this issue in the MTCS course and to deliver the message that a learning opportunity exists in each pupils' mistake (or misunderstanding). Several pedagogical tools for exposing learners' alternative conceptions are presented as well as three activities to be facilitated in the MTCS course.

6.1 Introduction

This chapter focuses on learners' alternative conceptions. "Alternative conceptions" is the customary term used nowadays for what was frequently called misconceptions in the past. The term *alternative conceptions* is used in this guide to highlight the legitimacy of these learners' conceptions, especially in the early learning stages.

Chapter 4 elaborates on mistakes and misconceptions. It emphasizes that mistakes should not necessarily be conceived negatively since they provide learners with the opportunity to correct their current knowledge and update their mental representation of the said topic or concept.

From a pedagogical perspective, since in many cases, misconceptions are consistent, systematic, are based on some modification of correct knowledge, and are stable in attempts to change them, teachers should look for misconceptions and help learners correct them by addressing their source. Further, the occurrences of learners' mistakes encourage teachers to use different tools to help learners improve their understanding.

In practice, a teacher cannot effectively deal with alternative conceptions without being aware of their mere existence, since alternative conceptions are not

© Springer-Verlag London Limited 2014
O. Hazzan et al., *Guide to Teaching Computer Science*,
DOI 10.1007/978-1-4471-6630-6_6

easily detected by conventional tests and other traditional assessment methods (see Chap. 10). More specifically, educators must be aware of their learners' ways of thinking and mental processes, must gain skills for uncovering alternative conceptions, and must recognize and use pedagogical tools to deal with these conceptions. In this spirit, knowledge of learners and their characteristics is one of Shulman's (1986) seven categories of the teacher knowledge base model.

Novice computer science educators, however, face difficulties in gaining the notion of alternative conceptions. Specifically, they face difficulties in:

1. Understanding how people do not understand topics which they conceive as trivial ones;
2. Lowering their level of understanding to that of a novice learner since their understanding of the subject area is usually more advanced;
3. "Getting into the head" of someone else (not just a pupil) because they have not yet the experience needed for performing such tasks.

These difficulties can be explained by Fuller's (1973) model, which distinguishes between three developmental stages of teacher experience: the self stage, the task stage, and the impact stage. While experienced teachers are usually at the third stage, and are concerned with issues related to *their pupils*, concentrating on what their pupils think, feel, and understand, novice teachers are mainly in the first (self) stage, dealing mainly with survival issues, such as how to keep their classes silent, how to complete teaching the curriculum on time, and how to make sure that they will be able to answer correctly their pupils' questions. Fuller found that the transition from the personal-centered stage to the pupil-centered stage is a function of experience; the more experienced a teacher is, a better chance exists for him or her to release themselves from survival issues and to concentrate on his or her pupils' thinking and behavior. Bents and Bents (1990) also recognize the transition from being a novice teacher to an expert teacher by replacing the attention given to teacher-centered issues to pupil-centered issues.

In order to prepare prospective computer science teachers toward the transition to the third stage, it is important to focus on learners' conceptions in the MTCS course. Within this context, one of the main messages that should be delivered to the students is that *a learning opportunity exists in every pupil's mistake* (or misunderstanding). However, in order to exhaust the pupil's learning abilities, it is necessary first, to understand the pupil's alternative conception and its source, and then, to use appropriate pedagogical tools to assist him or her improving their understanding.

It is important to remember, though, that it is impossible to prepare prospective computer science teachers for all the situations they will face in the future. Therefore, with respect to dealing with learners' alternative conceptions, the practical goals of the MTCS course are to increase the students' awareness with respect to:

1. The existence of a cognitive diversity in general and alternative conceptions in particular;
2. The potential contribution of positive attitude toward the phenomenon of alternative conceptions;
3. The stage in which they, as teachers, will transit from the self stage to the pupil-centered (impact) stage;
4. Availability of different pedagogical tools for identifying and dealing with alternative conceptions;
5. Common alternative conceptions described in the computer science education research literature (see Chap. 4).

6.2 Pedagogical Tools for Dealing with Alternative Conceptions

In what follows, we present several tools and strategies that can be used to expose and deal with alternative conceptions that prospective computer science teachers should be familiar with. We also explain how they can be addressed in the MTCS course.

Pupil–Teacher Interaction: A common assumption held by novice computer science teachers is that pupils' questions should be answered immediately and their problem should be solved right away. Therefore, many teachers tend to fix an error or a mistake in pupils' solutions or pupils' ways of thinking as soon as they recognize it. Teachers, however, should resist this tendency and listen to their pupils very carefully before they start seeking for the corrective teaching action. This approach is similar to the expected behavior from a doctor, i.e., not to suggest a medicine to a patient before he or she examines and understands the patient's problem and completes the diagnosis. In practice, as it turns out, in many cases, when a pupil asks a question, it is worthwhile just repeating the pupil's question, allowing the pupil to answer it first.[1]

In the MTCS course, it is recommended to discuss this approach with the students and to encourage them to start listening to their pupils from their first teaching experiences, e.g., in the practicum (see Chap. 13).

Diagnostic Exercises: A teacher can ask pupils a series of questions whose purpose is to reveal the pupils' alternative conceptions. Metaphorically, the teacher can develop a series of tests to expose pupils' cognitive bug. It is important to recognize the bug both from the pupil's and the teacher's perspectives; specifically, the pupil must recognize the bug in order to start debugging and modifying his or her cognitive model; the teacher should expose the pupil's bug in order to implement some pedagogical intervention.

---

[1] An illustrative metaphor compares the repetition of the question to bouncing a ball back to the pupil, letting the pupil be "in charge" of the ball (in our case, question).

An example of such a diagnostic exercise is included in Activity 44, presented below (Sect. 6.3). If time allows in the MTCS course, it is recommended to ask the students to prepare a similar set of exercises for other computer science topics. A related idea is addressed in Activity 28 (see Sect. 4.3).

Errors and Improving Understanding: The mere existence of mistakes should be acknowledged both in the computer science class and in the MTCS course. This can be done by showing learners' common errors and discussing them, talking about errors that learners exhibited in written tests, and most important, encouraging learners to conceive errors as unavoidable *positive* phenomena that comprise an opportunity to learn new ideas and improve current understanding. In some cases, it is recommended to present a common error (known from the computer science education research literature), even if none of the learners in the specific class did not suggest it, and to discuss it with the learners.

In many cases, teachers are exposed to learners' alternative conceptions through learners' wrong answers in written tests. In such cases, obviously, pupils describe neither their (alternative) conception as part of their wrong answer nor the way of thinking that led them to present the wrong answer. Therefore, the teacher should "get into" the pupil's head, recognize the concept or topic that the learner does not understand, try to imagine the pupil's conception of the said concept, and guess the pupil's intention in the (wrong) answer.

Activities 44, 45, and 46, to be facilitated in the MTCS course, practice this skill. Activity 44 is based on computer science education research (see also Chap. 4). Activities 45 and 46 focus on genuine answers given by pupils in a test and the alternative conceptions they expose: Activity 45 is based on the actual data taken from a written exam; Activity 46 examines a short interview conducted with a pupil who exhibited an alternative conception.

Activity 44: Exploration of a Computer Science Education Research Work on Learners' Understanding of Basic Computer Science Topics

As mentioned in Chap. 4, knowledge gained from computer science education research can enrich teachers' perspective with respect to pupils' difficulties, misconceptions, alternative conceptions, and different cognitive abilities and skills.

The general target of this activity is to present an example of a research work to the students that deals with learners' understanding of variables—one of the basic computer science concepts—including input, output, and assignments, and specifically, to focus on alternative conceptions, mistakes, and misconceptions related to variables.

The expected outcome of this activity is that in later stages, in the various assignments that the students will work on, they will exhibit a wide perspective on learners' mistakes. In other words, it is expected that in future activities, whether it is a lesson planning, a design of a specific learning activity or a tutoring activity, and so on, the students will focus on a variety of topics, such as learners' understanding, factors that may mislead learners, learning activities that can meet different learners' needs, and so on.

The activity is based on two stages: an individual or teamwork and a class discussion.

• › Stage A: Alternative conceptions about variables, individual or teamwork
The worksheet for the individual work or teamwork is presented in Table 6.1.
• › Stage B: Class discussion
The presentation of the research mentioned in stage A provides the instructor of the MTCS course with the opportunity to clarify to the students the difference between a mistake and a misconception. It also fosters a discussion about why a teacher should not be satisfied just by finding a learner's mistake and correcting it. The instructor of the MTCS encourages first, to initiate an in-depth discussion and investigate together with the students the reasons that caused the mistakes, and second, to recognize, once again, together with the students, ways that may guide learners confront their misconceptions and improve their understanding of the said concepts.

It is also important to connect this activity to the research in computer science education (Chap. 4) and emphasize how the familiarity with such research works can increase teachers' awareness to different learners' conceptions, difficulties, and experiences with computer science concepts.

Table 6.1 Worksheet about alternative conceptions related to variables

Worksheet
The following assignment is based on a questionnaire presented to novice Pascal learners as part of a research study (Haberman and Ben-David Kollikant 2001)
Learners were asked to write the output of each of the following programs.
(In Pascal: The input instruction is Read, the output instruction is Write, the assignment symbol is: =)

Program1	Program2	Program3	Program4
Read (A, B)	Read (A, B)	Read (A, B)	Read (A, B)
Read (B)	B: = 4	B: = 4	Read (B)
Write (A, B, B)	Write (A, B, B)	Write (A, B); Write (A)	Write (A, B)

Answer the following questions
For each of the 4 programs, write its output.
For each of the 4 programs, write one incorrect solution that you predict may be presented by a novice learner. Describe the learner's alternative conception that could lead him or her to present the incorrect answer.
In your opinion, is it possible that a student, who has an alternative conception related to the said concepts, will present a correct answer for these programs? If yes, present a different program that, in your opinion, will expose the alternative conception

Activity 45: Assessment of a Pupil's Answer in a Written Exam[2]
The target of this activity is to let the prospective computer science teachers experience one aspect of the assessment process (see also Chap. 10). For this purpose, the activity combines computer science topics, such as the role of arrays, and pedagogical ideas such as, how to evaluate an answer given by a pupil in an exam. Specifically, in the activity, the students analyze and evaluate a genuine answer that was given by a high school pupil in an authentic written exam.

• › Stage A: Checking a pupil's written answer, work in small teams
The students are presented with a question from a real test, alongside a pupil's authentic answer to that question. Working in small teams, the students are asked to read the answer, understand the pupil's intentions, explain the alternative conception that may lead the pupil to give this answer, and grade the answer. They are also asked to describe what they would ask the pupil if they had a chance to talk to this pupil. This issue is important to address since in school life teachers have many opportunities to talk to their pupils. The task is presented in Table 6.2.

• › Stage B: Class discussion
During the class discussion, it is recommended to follow the six tasks presented in the worksheet by the same order.

It is important to emphasize that the wrong answer (by itself) is not the most important issue in this activity and that it only serves as a trigger to discuss a variety of important pedagogical issues, such as, the understanding of pupil's intentions (Task 2) and talking to the pupil (Task 3).

With respect to Task 2, it is relevant to highlight the fact that though the revealing of the source of wrong answers is not an easy pedagogical activity, it is crucial if a teacher wishes to understand the mental model that lead pupils to present incorrect answers. It is, therefore, relevant to explore different strategies for recognizing pupils' intentions in writing exams.

Task 3 focuses on the assistance that a teacher may give to pupils. The students should realize that statements such as "your answer is wrong" or "you don't understand what we learned in class" are simply irrelevant since they do not help the pupil realize his or her wrong cognitive model and further, can decrease learners' motivation. Therefore, teachers should adopt a more clinical approach which confronts pupils with their answers and lead them to understand their mistakes. Activity 46 (see below) presents such a conversation in which a teacher uses concrete examples in order to lead the pupil to understand his wrong mental model.

The discussion of Tasks 4, 5, and 6 should lead to a variety of assessment-related conclusions (see Chap. 10). One of them highlights the fact that there is no single or unique way to evaluate pupils' answers in exams and that different approaches can be applied in the process of exam assessment.

---

[2] Based on Lapidot and Hazzan (2003).

These approaches, however, should make sense and be based on thoughtful pedagogical considerations, and when appropriate, should be explained to the pupils who take the exam. Another conclusion should increase students' awareness of the fact that sometimes they can, unintentionally, convey different messages (either explicit or implicit) to pupils through their comments written on pupils' exam papers.

Additional assessment-related issues that can be addressed in this context are: Should teachers award points for the correct parts of a pupil's wrong answer? Alternatively, when a mistake is observed, should teachers subtract points from the total points allocated for a specific question? What is the weight of each mistake? Should a teacher distinguish between logical mistakes, syntax mistakes, and programming style issues? Chapter 10 further elaborates on these questions.

Table 6.2 Worksheet about the assessment of a pupil's answer in a test

Worksheet: Assessment of an answer to a test
This worksheet presents a question that was given to high school pupils in a written exam after they had learned arrays, and a genuine answer given by one of the pupils. Work on the following tasks in teams.
1. Read the question and answer it. 2. Read the pupil's answer, try to understand the pupil's intentions, and explain why the pupil wrote that answer or, in other words, what could lead him or her to write this answer. 3. If you had a chance to talk to this pupil, how would you confront the pupil with the written answer? 4. Mark all the mistakes and problems you find in the answer. 5. Take a clean copy of the answer and regard it as if it was the pupil's exam notebook. Determine what to write to the pupil and how to present your comments. 6. Decide on the grade, out of 30 points, that indicates this answer's score.
The question
Write a program that calculates and prints the following data for a given natural (positive integer) number *n*, followed by *n* grades (0–100).
1. The average of all grades. 2. The number of grades which are higher than 55. 3. The lowest grade. 4. The number of grades that are higher than the average (found in #1).
The pupil's answer (indentations and comments were copied from the pupil's original notebook)

```
public static void main (String[] args)
{
        int
            w, z, x, a, n, f:
              w = 0; Z = 1000
                for(i = 1; i<= n; i++)              .......... 1,2,3
{                                                    number of grades

                a = input.nextInt();                the grades
                x = x+a;                             the sum
                }
              1)av = (x/n);                          average
                if (a>55) w = w+1
                if (a<z) z = a
                if (a>av) f = f+1
}

                System.out.println av; w; z; f

}
```

Teachers conduct clinical conversations with their pupils on a daily basis. In these talks, teachers improve their understanding of their pupils' difficulties and use this improved understanding to determine what pedagogical approach or a teaching action to apply. In many cases, these talks take place after a pupil claims a general statement that indicates some difficulty, such as "I don't understand," and the teacher's intention is to uncover the source of the pupil's difficulty.

Activity 46 highlights a different situation, in which a pupil is *not* aware at all about his wrong perception. In this case, the interview serves as a means to expose the pupil's understanding and as a teaching intervention that guides the pupil to find the correct answer. Based on this activity, the students in the MTCS course should increase their awareness of the fact that sometimes, instead of telling pupils that their answer is wrong, a series of questions can lead pupils not only to derive this conclusion by themselves, but also to correct their answers.

Activity 46: A Clinical Conversation with a Pupil as a Means to Reveal Alternative Conceptions

- Stage A: A clinical conversation with a pupil, work in small teams

The students in the MTCS course are given a transcription of a conversation between a researcher and a pupil, David. At several points they are asked to stop reading and answer several questions. The activity is presented in Table 6.3.

- Stage B: Class discussion

After the students finish their work, it is recommended to facilitate a class discussion which concentrates on the last two parts of the worksheet: David's conceptions and planning an optional series of questions for David. It is also optional to split the discussion into several parts, i.e., a short discussion after each part of the worksheet.

It is important to deliver the message that the researcher could ask another question in this conversation which might lead to either similar or different pupil's conclusions. Specifically, while the researcher's questions related to the fact that David assumes that the variable a contains many values, other problems in David's answer could be addressed, such as the fact that he used random numbers instead of reading the input, and the fact that he did not use any repetitive (loop) instruction. The discussion may focus on specific questions that may lead David to realize his difficulties with respect to these issues.

Table 6.3 A worksheet on a clinical conversation with a pupil

Worksheet: A "clinical conversation" with a pupil

David is a high school pupil who just finished writing an exam. The interview with him took place just after the exam. He was very happy and had a good feeling about his answers. He was asked to present his solution to one of the questions.

In what follows, the question is presented. Then, the interview transcription is laid out.

Please read the transcription and answer the questions presented after each part of the interview.

The exam question:
Write a program that prints the positive numbers from any given 35 numbers.
1. Read the question and answer it.

Interview – Part 1

(R – represents the researcher, D – represents David)
D: The test was great, I did it very well.
R: How did you solve question 9?
D: I wrote [David writes his answer on a paper:]

```
public static void main (String[]args) {
        int a;
        a=(int) (Math.random() * 36);
        System.out.println(a);
        if (a>0) System.out.println(a);
}
```

2. Before going on reading the interview, try to understand David's intention in his answer.
3. As a teacher, what would you like to ask David now?

Interview – Part 2

R: Please explain why you wrote here Math.random() * 36? [the researcher points at the assignment]
D: I did 36 because you need 35 numbers and it [the computer] doesn't give the last one, so you need another one [number].
R: How many numbers do you think there are now in variable a?
D: 35 numbers, like I did here [points at the assignment] and here I print them one after the other [points at the assignment before the if statement].
4. Explain David's conceptions. To what concepts are these conceptions related?
5. As a teacher, what would you like to ask David now?

Interview – Part 3

R: Let's try to run the program together. Please tell me which numbers you think will be in a.
D: Ah ... for example 1 7 9 3 [the researcher writes the numbers on a paper, one below the other]
R: Is −3 possible?
D: Yes, Oh, I didn't do it well. I have a mistake with−3.
R: Never mind, let's continue. Is 100 possible?
D: Yes
R: Let's put aside for just a moment what you wrote. Please look at the following
    instructions:
    a = 3;
    a = 7;
    System.out.println (a);
    What do you think will be printed?
D: [answers immediately] 10
R: Why 10?
D: Oh, I was wrong. It will print only 7. I have to add in my solution another [instruction of] System.out.println(a) [adds it in the original program before the if instruction, see below]

```
public static void main (String[]args) {
int a;
a=(int)(Math.random() * 36);
System.out.println(a);
System.out.println(a);          ⇐the new line

if (a>0) System.out.println (a)

}
```

[At this stage the interview stopped due to interruptions from other pupils in the room].

6. Did the interview improve your understanding about David's conceptions?

7. Would you ask David different questions than the researcher did? Please prepare a series of questions to ask David.

References

Bents M, Bents R (1990) Perceptions of good teaching among novice, advanced beginner and expert teachers. American Educational Research Association, Boston

Fuller FF (1973) Teacher education and the psychology of behavior change: a conceptualization of the process of affective change of preservice teachers. Annual Meetings of the American Psychological Association, Montreal, Canada

Haberman B, Ben-David Kollikant Y (2001) Activating "black boxes" instead of opening "zippers"—A method of teaching novices basic CS concepts. SIGCSE Bull 33(3):41–44

Lapidot T, Hazzan O (2003) Methods of Teaching Computer Science course for prospective teachers. Inroads—SIGCSE Bull 35(4):29–34

Shulman LS (1986) Those who understand: knowledge growth in teaching. Educ Res 15(2):4–14

Teaching Methods in Computer Science Education

7

Abstract

This chapter presents active-learning-based teaching methods that computer science educators can employ in the classroom. The purpose of this chapter is first, to let the students in the Methods of Teaching Computer Science (MTCS) course experience a variety of teaching methods before becoming computer science teachers; second, to discuss, together with the students, the advantages and disadvantages of these teaching methods; and third, to demonstrate high school teaching situations in which it is appropriate to employ these teaching methods. Within this chapter, we discuss (a) pedagogical tools: games, the CS-unplugged approach, rich tasks, concept maps, classification, and metaphors; (b) different forms of class organization; and (c) mentoring software project development.

7.1 Introduction

This chapter presents active-learning-based teaching methods that computer science educators can employ in their classroom (see Chap. 2). We focus on nontraditional methods in order to encourage computer science educators to employ these methods in their classes.

In the context of the Methods of Teaching Computer Science (MTCS) course, the purpose of this chapter is first, to let the students experience a variety of teaching methods before becoming computer science teachers; second, to discuss advantages and disadvantages of these teaching methods; third, to demonstrate high school teaching situations in which it is appropriate to employ these teaching methods; and finally, to vary the teaching methods employed in the course. Since, in most cases, the activities carried out in the MTCS course focus on a specific computer science topic, they also provide the prospective computer science teachers with additional opportunity to improve their own understanding of computer science concepts.

© Springer-Verlag London Limited 2014
O. Hazzan et al., *Guide to Teaching Computer Science*,
DOI 10.1007/978-1-4471-6630-6_7

The teaching methods presented in this chapter are pedagogical tools (Sect. 7.2), different forms of class organization (Sect. 7.3), and mentoring software project development (Sect. 7.4).

For each teaching method, we first outline its meaning, target, and importance in computer science education. Then, we present active-learning-based activities to be facilitated in the MTCS course. The actual facilitation of these activities in the MTCS course is important since the students should sense how their future high school pupils will feel when they, as computer science teachers, will employ these teaching methods in their high school classes.

We note that several of the teaching methods presented in this chapter are illustrated also in other chapters of this guide when the focus is placed on another pedagogical topic.

7.2   Pedagogical Tools

In this section, we review the following pedagogical tools:

- Pedagogical games (Sect. 7.2.1)
- The CS-unplugged approach (Sect. 7.2.2)
- Rich tasks (Sect. 7.2.3)
- Concept maps (Sect. 7.2.4)
- Classification (Sect. 7.2.5)
- Metaphors (Sect. 7.2.6)

7.2.1   Pedagogical Games

Games have a great pedagogical potential. A well-planned game enables to learn new concepts in an alternative class atmosphere, involves social interaction, introduces a change in the teaching method, and is a kind of activity that all students are good at. Games, as other pedagogical tools, may also have disadvantages, such as the chaos that a game may cause in the class, dominant students' takeover of the game process, learners' disagreement to participate in an activity they conceive childish, and teacher's inability to control the class as they do in traditional teaching settings. Table 7.1 summarizes some advantages and disadvantages of games from pedagogical, social, and emotional perspectives.

We note that we discuss here neither computer games nor game programming (e.g., as in Lakanen et al. 2014), which has become popular recently also in computer science education (see Activity 48); rather, we focus on social games that aim at teaching computer science ideas, which can be played either with or without computers. We note also that the use of games as a pedagogical tool is mentioned also in the next section, which focuses on the CS-unplugged approach, which is a computer science teaching method which is largely based on noncomputerized games and activities.

Table 7.1 Advantages and disadvantages of games from pedagogical, social, and emotional perspectives

	Pedagogical	Social	Emotional
Advantages	Based on active learning	Enables all students to participate	Breaks the routine
		Enhances interactivity (if played with more than one player)	
		Most games are competitive and can increase motivation	
Disadvantages	May distract learners' attention from the intended computer science content	Dominant students may control the game process	If a player loses a game, it may influence his or her feeling regarding computer science learning
	Teachers' conception that time is wasted and that no meaningful learning takes place while playing	May cause chaos in the class	
		Learners' disagreement to participate in an activity they conceive childish	
		Most games are competitive and may distract some learners	

Activity 47 presents a lesson (or two—pending on the available time) of the MTCS course that focuses on how to use games in computer science education.

> **Activity 47: Pedagogical Examination of Games**
>
> - Stage A: Playing a game
>   In the spirit of the active-learning-based teaching model (see Chap. 2), the lesson starts with playing a pedagogical game. The Conditional-Statement-Bingo (Hebetim 1995), presented in Table 7.2, is one optional game. Such experience may help the students sense the potential use of games in computer science education and may expand their creative ideas for additional games that they will be asked to design in a later stage.
> - Stage B: Class discussion
>   After the students had felt the pedagogical potential of games in the context of computer science learning (Stage A), they are presented with several questions (see Table 7.3) in order to initiate a discussion about the topic. The questions can be presented orally or as a worksheet on which the students work as a preparation toward the class discussion. In the discussion, the instructor can choose the questions to focus on according to the lesson's flow.
> - Stage C: Game design
>   Based on the discussion that took place in Stage B, the instructor, together with the students, selects one kind of game (e.g., an outdoor game) and

the students are asked to work in teams and to design such a game. It is recommended to ask all the teams to design a game on the same computer science topic (according to your choice). They should also be asked to explain the pedagogical guidelines they followed.

- Stage D: Presentation and class discussion
  Each team presents the game it designed, together with the pedagogical guidelines it followed. With respect to each game, it is important to first address its targets, and second—its potential contribution to the learning of the said computer science topic.
- Stage E: Design and construct a game
  The students are asked to design and construct at home another game, of a different kind than the one designed in class, on any computer science topic, according to their choice. In this case, they are asked also to construct the game, that is, to submit a self-contained work that allows computer science learners start playing the game.
  Again, they should also explain the target and the pedagogical guidelines they followed in the design and construction process of the game.
- Stage F: Playing the games
  In the lesson, in which the students submit the games they designed and constructed at home, it is recommended to let each of them present his or her game shortly and then, to select one or two games and let the students play it. After they play the game, it is important to reflect on their experience while playing, addressing cognitive, social, and emotional aspects.
- Stage G: Summary
  In summary, it is important to address the advantages and disadvantages of incorporating games in computer science education and to highlight these advantages and disadvantages from pedagogical, social, and emotional aspects. Table 7.1 can guide the presentation of such a summary. Needless to say, the working assumption is that the games we refer to are well designed and pedagogical thought has been invested in their development. One of the important conclusions of this lesson should deliver the message that though games are effective teaching tool, they should be used thoughtfully based on careful examination of their suitability to each class.

Table 7.2 The Conditional-Statement-Bingo game. (©Hebetim 1995)

The Conditional-Statement-Bingo game
In the regular Bingo game, each player holds a board with numbers. The game coordinator raffles a number and announces it to all players. Players, who have this number on their boards, mark the cells with the raffled number. The winner is the first player who successfully marks all the cells on his or her board
The Conditional-Statement-Bingo game is based on the same idea. It can be played with the whole class or in small groups. The coordinator can be one of the learners
The Game—overview

Table 7.2 (continued)

The Conditional-Statement-Bingo game
Each board has nine cells and each cell includes a valid conditional statement (see Fig. 7.1). In each such statement, the instruction System.out.println ("Bingo") appears in one of the statement branches. That is, for each *if* statement, the word Bingo is printed at one of its branches, according to the variables' values. For example, if c's value is bigger than 80, Bingo is printed in the following case: if (c > 80) System.out.println ("Bingo"); and the cell is marked
Each player's target is to mark all the cells on his or her board, according to the following rules
The Game—Rules
The game coordinator shuffles the cards and puts them in a deck with their face down. The coordinator takes one card and reads its content. The cards include assignments to variables *a*, *b*, or *c*, for example: $a = -7, b = 2 \times (10 + 15)$
After the coordinator reads the assignment statement, each player checks what cells in his or her board should be marked; that is, a cell can be marked, if the System.out.println ("Bingo"); statement is executed according to the announced values
The game ends when the card deck is empty or when one of the players marked all the cells in his or her board
The game can be prepared in different levels of conditional statements difficulty, according to the class' level

Table 7.3 Questions about games as a pedagogical tool is computer science education

Questions about games as a pedagogical tool in computer science education
What kinds of social games are you familiar with? (e.g., indoor/outdoor games; games with a different number of players, etc.)
In what cases does it make sense to use social games as a pedagogical tool in computer science education?
How can games enhance computer science learning?
Do games fit a specific population of computer science learners? Explain your opinion
What topics are specifically suitable to be learned by games?
What disadvantages does the teaching with games have?
Can a teacher evaluate learning processes during game playing?
In what frequency should a teacher use a game as a pedagogical tool?
What is the teacher's role during the game?

if (a < 5)
  System.out.println("Bingo");
	if (b > 2)
	  System.out.println("Hello");
	else
	  System.out.println("Bingo");
		if (c == 50)
		  System.out.println("Bingo");
		else
		  System.out.println("Hello");

if (a > -10)
  System.out.println("Hello");
else
  System.out.println("Bingo");
	if (b < 34)
	  System.out.println("Bingo");
		if (c > 0)
		  System.out.println("Big");
		else
		  System.out.println("Bingo");

if (a == 10)
  System.out.println("Hello");
else
  System.out.println("Error");
	if (b == 120)
	  System.out.println("Bingo");
	else
	  System.out.println("No");
		if (c > 80)
		  System.out.println("Bingo");

Fig. 7.1 An example for a board of the Conditional-Statement-Bingo game

Activity 48: Educational Usage of Games in Computer Science Education, Group work
As mentioned, in recent years we see usages of computer games in educational context as well. This activity, which is carried out by teams, exposes the prospective computer science teachers to this aspect of game usages in computer science education.

The students are asked to find at least three educational programs in computer science that are based on computer game development. Their task is to present their review in a 15 min presentation. The presentation should include at least 5 min of gaming.

7.2.2 The CS-Unplugged Approach

The CS-unplugged approach delivers the message that "Computer Science isn't really about computers at all!" In this spirit, different groups around the world developed a series of learning activities that aim at teaching a variety of computer science concepts, such as, binary numbers, sorting algorithms, data structures and data compression, without connecting them directly either to computers or programming. One of the main resources for such activities is the extensive free collection developed by the Computer Science Unplugged team.[1] Additional groups around the world adapted this approach for different populations. In all cases, the computer science concepts are presented through engaging activities and puzzles by using cards, crayons, and active playing.

The purpose of the Activity 49, that focuses on the CS-unplugged approach, is to increase students' awareness to several facts related to the use of computers in computer science teaching, such as, differences between teaching computer science with and without computers, the variety of computer science teaching methods, strategies for the selection of a suitable teaching method for the teaching of a specific computer science topic to a specific population, and the option to teach computer science concepts independently of technology. These topics should be shortly discussed in the MTCS course in between the stages of the activity.

Activity 49: Pedagogical Examination of the CS-Unplugged Approach

- Stage A: Experience a CS-unplugged activity
  The instructor selects one of the CS-unplugged activities and invites the students to experience it before addressing it from a pedagogical perspective.
- Stage B: Exploration of the CS-unplugged approach
  The students are asked to look at the CS-unplugged website, to select two topics from this website and analyze, from a pedagogical perspective, the suggested activities for the teaching of the said computer science topics.

---

[1] See http://www.csunplugged.org/.

- Stage C: Design of a CS-unplugged activity, work in pairs
  The students work in pairs, and for one topic on which they worked in Stage B they are asked to design another activity that teaches it without computers.After the students design their activities, they present them in front of the class, along with the pedagogical guidelines they followed in the design process.Then, a summary should be presented that organizes the main issues that came up during students' work. In this summary, it is important to highlight the fact that both teaching approaches (with and without computers) have advantages and disadvantages, and that each approach can be used in different teaching situations.
- Stage D: The CS-unplugged approach and other computer science teaching methods, homework
  First, the students are asked to design an activity with a computer that teaches the topic they selected in Stage C. Second, they are asked to analyze similarities and differences between the two activities they designed: the one that uses computers and the one that does not use computers, and to analyze advantages and disadvantages of the teaching approach that each activity represents.

7.2.3 Rich Tasks[2]

Rich tasks (Lapidot and Levy 1993; Levy and Lapidot 1997) are programming exercises that (a) can be solved in a variety of ways, when each solution elicits and promotes a discussion about one or more major computer science ideas; and (b) can be solved within the duration of one lesson based on learners' current knowledge. Rich tasks can focus on any computer science subject.

The importance of rich tasks stems from the fact that in the learning/teaching processes of programming, it is sometimes easier to focus on technical aspects. Since this temptation is especially relevant for novice computer science teachers, and since rich tasks illustrate how even the teaching of basic computer science concepts can be designed in a way that highlights computer science "big ideas," it is important to address this pedagogical tool in the MTCS course.

The potential contribution of rich tasks to the learning of computer science is evident; specifically, since rich tasks highlight computer science concepts within a programming context, learners acquire also more abstract knowledge, in addition to computing-oriented knowledge. Clearly, the concept of a rich task can be applied also to nonprogramming tasks; however, we highlight the notion of a rich task in the context of programming to increase the prospective computer science teachers' attention to the potential contribution of programming tasks also to learners' conceptual knowledge (in addition to technical knowledge, on which, as mentioned, there is sometimes a tendency to focus in such cases).

---

[2] ©Hebetim (2005).

We highlight several pedagogical aspects related to rich tasks:

- A rich task is a well-selected programming task that has advantages from both cognitive and pedagogical perspectives. For example, from a cognitive perspective, it elicits learners' thinking on a higher level of abstraction; from a pedagogical perspective, a rich task demonstrates an active learning teaching approach which encourages all learners to express their creativity and promotes social interaction.
- The main purpose of rich tasks is to highlight computer science ideas (see, e.g., the list presented at Stage A of Activity 50); the solutions themselves and the technical details should get less attention.
- The different solutions of rich tasks can be presented in the class in different orders. In general, the presentation order should be determined based on the pedagogical targets of the computer science educator, as well as on learners' background.
- When a lesson is planned around a rich task, it is important to take into consideration that learners tend to focus on technical aspects, rather than on conceptual notions, and to be satisfied when they get a working solution. Therefore, conceptual computer science ideas should be addressed only after the teacher verifies that all the pupils in the class solved the task (i.e., wrote a running program), and only based on this realization, to move on to the more conceptual discussion. In fact, as hinted, it is important to realize that also for teachers/instructors it is sometimes easier to focus on technical issues. However, as has been mentioned previously in this guide, the softer computer science concepts should not be neglected in computer science education (see Chap. 3).

In the MTCS course, in addition to the introduction of this pedagogical tool to the students, it is recommended to let the students experience working on rich tasks, to examine this teaching approach from a pedagogical perspective, and to construct a rich task. Activity 50 illustrates how this multifaceted perspective can be achieved by one specific rich task.

Activity 50: Pedagogical Examination of Rich Tasks
This task is intended for novice learners. The same principles, however, can be applied for advanced computer science concepts and learners. We chose to discuss this particular rich task in the MTCS course to demonstrate that many different solutions exist even for a relatively simple task when each solution highlights (a) different important computer science concept(s).

- Stage A: Solving a rich task, individual work
  The students are asked to work on the task, presented in Table 7.4.
  While the learners are working (with or without computers) on this task, the educator walks around in the class in order to collect different solutions. He or she can also talk to the learners, encourage them to develop more than one solution, and locate misconceptions and other powerful computer science and pedagogical ideas to be addressed later in the lesson.

In the next stage, we see that this task has the potential to raise discussions about a wide variety of computer science topics, such as:

- Abstraction
- Programming style
- Conditional statement, different brunching strategies, linearity versus nesting
- Imperative versus functional approach
- Efficiency
- Readability
- Procedures/methods, functions
- Meaningful names
- Boolean expressions and functions, the flag metaphor
- Data types
- Parameters
- Sophisticated (tricky) solutions

As can be easily observed, these concepts are varied and highlight both rigid and soft aspects of computer science; therefore, such tasks, though look relatively simple, are called rich tasks.

- Stage B: Presentation of learners' solutions
  After the learners worked on the task, the solutions are presented by the instructor and discussed with the whole class. The instructor, however, does not set the order of presentation randomly or according to the learners' desire to present their solution; alternatively, in order to exhaust the potential pedagogical contribution of this task, the instructor selects very carefully the order of presentations (and even prepares it a priori). The solutions can be presented by, for example, their readability level, their algorithm complexity, their sophistication level, or the level of the computer science ideas they demonstrate. For each solution, it is recommended to highlight what computer science ideas it illustrates.
  Table 7.5 presents possible solutions for this task. If time permits, it is suggested to let the learners capture the essence of each solution by giving it a title.
- Stage C: Class discussion
  In the discussion, it is important to highlight the messages emphasized in the introduction of the concept of rich task.
- Stage D: Construction of a rich task, homework
  As homework, the learners are asked to construct another rich task and to explain the guidelines they followed. It is important to note that it is not a simple task to construct a rich task. In fact, the construction of a rich task is a kind of a rich task by itself since it requires the consideration of different options and the exploration of the computer science ideas that can be expressed by the possible different solutions of the task.
  We present one additional example of a rich task: Write a function that returns the maximum of four given numbers.

Table 7.4 A rich task

Task
Write a method that checks whether a given date is valid
The method checks whether three given integers (a day, a month, and a year) can represent a valid date in the twenty-first century
Assume that each month has 30 days

Table 7.5 Possible solutions of the rich task

Possible solutions of the rich task

Solution 1

public static void date (int year, int month, int day) {if (year>1999 && year<2100) {if (month>0 && month<13) {if (day>0 && day<31) System.out.println ("This is a legal date"); else System.out.println ("Day is NOT legal");} else System.out.println ("Month is NOT legal");} else System.out.println ("Year is NOT legal");}

Solution 2

```
public static void date (int year, int month, int day) {
    if (year>1999 && year<2100)
        if (month>0 && month<13)
            if (day>0 && day<31)
                System.out.println ("This is a legal date");
            else
                System.out.println ("Day is NOT legal");
        else
            System.out.println ("Month is NOT legal");
    else
        System.out.println ("Year is NOT legal");
}
```

Solution 3

```
public static void date (int year, int month, int day) {
    if (year<2000 || year>2099)
        System.out.println ("Year is NOT legal");
    else if (month<1 || month>12)
            System.out.println ("Month is NOT legal");
        else if (day<1 || day>30)
                System.out.println ("Day is NOT legal");
            else
                System.out.println ("This is a legal date");
}
```

Solution 4

```
public static void date (int year, int month, int day) {
        if (validYear (year))
                checkMonthAndDay (month, day);
        else
                System.out.println ("Year is NOT legal");
}
public static boolean validYear (int year) {
        return (year>1899) && (year<2100);
}
public static boolean checkMonthAndDay (int month, int day) {...}
```

Solution 5 (in this solution, we can ignore the location of the solution in the program and the source of the variables' values)

```
        if (year<2000 || year>2099)
```

Table 7.5 (continued)

```
            System.outprintln ("Year is NOT legal");
      if (month< 1 11 month> 12)
            System.out.println ("Month is NOT legal");
      if (day<1 || day>30)
            System.out.println ("Day is NOT legal");
      System.out.println ("This is a legal date");
```

Solution 6

```
public class S6 {
      public static boolean wrongDate = false;
      public static void main (String[] args) {
      // input statements for year, month, day
            checkYear (year):
            checkMonth (month);
            checkDay (day);
            if (wrongDate)
                    System.out.println ("This is NOT a legal date");
            else
                    System.out.println ("This is a legal date");
      }
      public static void checkYear (int year) {
          if (year <2000 || year >2099) {
             wrongDate - true;
             System.outprintln ("Year is NOT legal");
          }
      }
      public static void checkMonth (int month) {
          if (month < 1 || month >12) {
             wrongDate = true;
             System.out.println ("Month is NOT legal");
          }
      }
      public static void checkDay (int day) {
          if (day<1 || day>30) {
             wrongDate - true;
             System.out.println ("Day is NOT legal");
          }
      }
   }
}
```

Solution 7

```
public class S7 {
          public static boolean wrong = false;
          public static void main(String[] args) {
          //input statements for year, month, day
             check (year, 2009, 2099, "Year");
```

Table 7.5 (continued)

```
            check (month, 1, 12, "Month");
            check (day, 1, 30, "Day");
            if (wrong)
                System.out.println (:This is NOT a legal date");
            else
                System.out.println ("This is a legal date");
        }
        public static void check(int number, int low, int high, String message) {
            if (number<low || number > high) {
                wrong = true;
                System.out.println (message +"is NOT legal");
            }
        }
    }
}
```

Solution 8 (in this solution, we can ignore the location of the solution in the program and the source of the variables' values)

```
        String[] message = new String[2];
        boolean legalYear, legalMonth, legalDay, legalDate;
        int isLegalDate = 0;

        message [0] = "This is NOT a legal date";
        message [1] = "This is a legal date";

        legalYear = (year>1999) && (year<2100);
        legalMonth = (month>0) && (month<13);
        legalDay = legalYear && legalMonth && legalDay;
        if (legalDate)
            isLegalDate = 1;
        System.out.println (message [isLegalDate]);
```

7.2.4 Concept Maps

According to Novak and Cañas (2008), "concept maps are graphical tools for organizing and representing knowledge. They include concepts, usually enclosed in circles or boxes of some type, and relationships between concepts indicated by a connecting line linking two concepts. Words on the line, referred to as linking words or linking phrases, specify the relationship between the two concepts."

For illustration, we present a concept map that partially represents the MTCS course and the connections between some of the topics it addresses (see Fig. 7.2).[3,4]

---

[3] Due to space limitations, only part of the topics addressed in the MTCS course are included in the map.

[4] Sometimes, different kinds of shapes and arrows are used to indicate different kinds of concepts and different types of relations among them. For the sake of simplicity, we decided to use the same shape (rectangle) for all concepts and the same kind of arrow for all types of relationships between them.

It shows, for example, that the MTCS course addresses research in computer science education that examines teaching methods and that teaching methods may apply active-based learning; it also reflects the fact that the theme of evaluation is addressed in the course with respect to the assessment of learners' conception. In this spirit, throughout this guide, we try to reflect these connections by mentioning cross-references to different chapters.

Since, as is illustrated in Fig. 7.2, concept complexity is reflected in its concept map, in most cases, a concept map is constructed gradually: first, core concepts related to the said topic are identified and located on the map, and then, relations between these concepts are identified, named, and located. Following the creation of the first version of the concept map, learners can update/change the map, either by reshaping its form or by adding/removing specific concepts and/or relationships.

When examined from a constructivist perspective (see Chap. 2), it is clear to realize the potential contribution of the construction process of concept maps to learners' understanding of the concept for which the map is constructed. While building a concept map, learners work with concepts and relationships between them, manipulating them both mentally (in their mind) and physically (in the map); thus, in parallel to the gradual actual construction process of the map, they construct their mental image of the topic represented by the map. For additional details about the theory underlying concept maps and how to construct and use them, see Novak and Cañas (2008).

Concept maps can be used for different pedagogical purposes, such as a summary of a topic after its main concepts were taught, identification of learners' alternative conceptions (see Chap. 6), and evaluation of learners' understanding of the said topic (see Chap. 10). In practice, in the classroom, learners can be asked to build a concept map from scratch; in other cases, an empty map can be given to the learners when they are asked to locate in it concepts and relations presented in a given list.

Concept maps are relevant for computer science teaching and learning processes in general and for the MTCS course in particular for several reasons. We mention two: First, concept maps can be used for different pedagogical purposes; second, the nature of computer science and its many facets represent different abstraction levels and interconnections among them.

Accordingly, Activity 51 that focuses on concept maps can be facilitated in the MTCS course after the idea of concept map is briefly introduced. This short introduction can be accompanied with a relatively simple example, not necessarily taken from computer science (the Web offers plenty of examples of concept maps), and should present the main guidelines for how to construct concept maps, without delving yet into their pedagogical usages.

Another option is to use concept maps as a mid-term task or as the summary task of the semester, in which the students are asked to construct a concept map which represents what they have learned so far in the course. Not only does such a task foster students' reflection about what they have learned so far in the course but it also encourages them to organize what they have learned so far in one framework that binds computer science topics with pedagogical and cognitive aspects of computer science education.

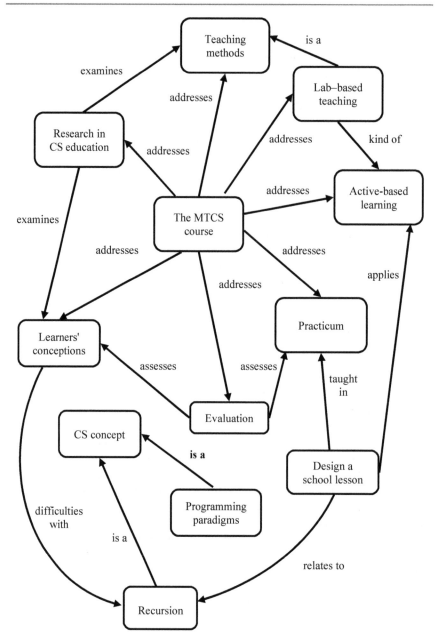

Fig. 7.2 Representation of the Methods of Teaching Computer Science (MTCS) course by a concept map

Activity 51: Pedagogical Examination of Concept Maps

- Stage A: Concept map construction, group work
  In order to discuss meaningfully the use of concept maps in computer science education, the students should experience first the construction process of a concept map. For this purpose, the students are asked to work in groups and to construct a concept map for a given computer science topic. To exploit the benefits of concept maps, it is recommended to select a topic in which both its rigid and soft aspects are dominant, for example, variables or complexity. In addition, each group should select one of its members to document the group's work during the construction process. This documentation is used later (in Stage C) in the class discussion that focuses on how to use concept maps in computer science classrooms.
- Stage B: Concept map evaluation, group work
  This stage focuses on the evaluation of concept maps.
  First, the class discusses evaluation rules for concept maps. The instructor should navigate this discussion in a way that, at its end, an evaluation scheme for concept maps that reflects learners' understanding of the said topic is formulated.
  Then, the groups rotate the concept maps they constructed in Stage A, and each group evaluates the concept map it received according to the evaluation scheme that has just been defined by the class.
- Stage C: Class discussion
  In this class discussion, the two first stages are summarized. This discussion should address both cognitive aspects (expressed mainly in Stage A) and pedagogical aspects (expressed mainly in Stage B) of concept maps.
  With respect to the cognitive aspect, it is important to highlight 1. the role and contribution of active learning to a meaningful mental construction of the learned topic and 2. the fact that the construction process of a concept map is a dynamic process, in which students reshape and reorganize both the concept map as well as their mental image of the said topic.
  The team members who documented the group work in Stage A can present their summaries at this stage.
  With respect to the pedagogical aspect, it is important to highlight the fact that the evaluation of concept maps should reflect their essence, and therefore, guidelines, such as the following ones should be included in any evaluation scheme:
  - The main concepts related to the said topic should be included in the map.
  - The relationships between the concepts should be well defined.
  In this class discussion, it is important to review also other pedagogical usages of concept maps.
  If times allows (or as a homework assignment), it is recommended to let the students construct a concept map for another computer science topic; this time, when they are aware of cognitive and pedagogical aspects of concept maps.

7.2.5   Classification of Objects and Phenomena from Life

In Sect. 3.7.2, we presented a classification task in the context of computer science soft ideas (see Activity 17). In this section, we expand the discussion about classification tasks in general, and about the use of classification tasks in the MTCS course in particular.

Specifically, we introduce classification as a teaching method in computer science education. We concentrate on the classification of *objects and phenomena from real life* to support and guide the learners' mental construction of the discussed computer science concepts. The working assumption is that familiar objects and phenomena enable learners to concretize computer science concepts which, in later stages, as is illustrated below, are conceptualized on a higher level of abstraction. This kind of task relies on the constructivist approach (see Chap. 2), according to which learners construct their knowledge by a meaningful engagement and active learning in an environment designed for the specific teaching target.

In the MTCS course, this kind of tasks form a basis for a discussion about the importance of allowing novices to work with different representations of computer science concepts in order to support their gradual mental construction. The following classification activity (Activity 52) illustrates this notion with respect to *control structures*.

Activity 52: Pedagogical Examination of Classification

- Stage A: Classification activity, group work
  The activity begins by presenting the students with different objects and phenomena taken from various sources: music, literature, transportation, newspapers, and so on (see Fig. 7.3). As can be seen, all the examples are taken from everyday life and experiences, and none of them is taken from the computing world.
  The students are not informed what the topic of the page is. The students are asked to work in groups and to classify these instances according to different criteria according to their own choice. Clearly, there is no correct classification. The students are also asked (a) to expand their classified sets by adding new instances to each set, (b) to give a title to each set, and (c) to suggest a title for the whole page. The specific instructions are presented in Table 7.6.
- Stage B: Class discussion
  After the groups have worked on their classification, each group shares its categorization with the rest of the class. This process can be performed in different forms. Here are several suggested forms:
  1. A group presents the instances of a specific set and the whole class should guess its classification criterion.

2. A group presents its additional new instance for a specific group and the class should guess which of the presented instances in the worksheet belong to this set.

3. A group presents its title for one of the sets and the whole class should guess which instances belong to that set.

During the discussion, the instructor should encourage a dynamic discourse and introduce generalizations and formal terminology with respect to *control structures*. Learners in general and the prospective computer science teachers in the MTCS course in particular, are often exposed in this discussion to new concepts and to different ideas and perspectives. This exposure, in turn, encourages them to reconsider their previous perspective at the topic (control structures, in this case) and to modify it, if needed. The discussion can end with a summary of the main concepts related to control structures that have been introduced in this activity.

• Stage C: Construction of a classification activity, homework
After the classification activity is facilitated with respect to control structures, it is recommended to ask the students to construct a similar activity with respect to another computer science concept (as a course activity or as homework). As a preparation for this activity, it is important to discuss with the students the essence of the concepts for which such an activity fits and to define criteria for the selected items. We mention here that recursion (see Chap. 12 for a similar activity), abstraction, and data structures are appropriate candidates for this purpose, and that the items selected to be included in the worksheet should clearly (and not in a hidden and sophisticated manner) reflect the illustrated concept. While preparing such an activity, it should be remembered that such classification worksheets are constructed for novice learners who are not familiar yet with the concept on which the worksheet concentrates, and therefore, the image or text should direct them clearly and safely in the identification process of the properties of the said concept.

We end this section by noting that although the classification activity is usually designed as an opening trigger in the learning process of a new concept, it is also possible to use it as a summary activity. In this case, all the stages and instructions will be facilitated in a similar way; learners' behavior, however, will be naturally different since they will be familiar with the main concepts/aspects of the said topic.

7.2.6   Metaphors

Metaphors are used in order to understand and experience one specific thing by using an analogy to another thing, usually a familiar concept (Lakoff and Johnson 1980). Many metaphors are used in computer science on a regular basis; just

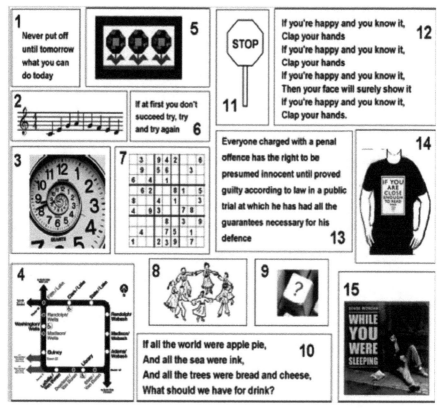

Fig. 7.3 Classification worksheet for control structures[5] (© Migvan—Research and Development in Computer Science Education, Department of Education in Science and Technology, Technion—Israel Institute of Technology)

think about pointers, a menu, windows, a mouse, a tree, and the computer memory. Therefore, and since the use of metaphors is also a powerful pedagogical tool, computer science teachers should recognize common metaphors used in the field, be aware of their importance, and learn how to use them meaningfully in learning and teaching processes.

---

[5] Resources of the items included in the control structure classification worksheet (Fig. 7.3): Item #3 http://homes.bio.psu.edu/people/faculty/bshapiro/research.html. Item #4 http://www.chicago-l. org/operations/lines/loop.html. Item#5 http://www.quiltdesignnw.com/Q132-SimplySunny-Easy-Kaleidoscope-Flower- quilt-pattern.htm. Item #8 http://www.junewatts.com/wwwcd.htm. Item #9 http://vanelsas.wordpress.com/2009/04/03/questions/. Items #10 and #12 Lyrics from Mother Goose. Item #13 Article 11 of the Universal Declaration of Human Rights of the General Assembly of the United Nations, formulated on December 10, 1948: http://www.un.org/en/documents/udhr/. Item #14 http://www.redbubble.com/people/taniadonald/t-shirts/1340952-3-if-you-are-close-enough-to-read-this-you-can-blow-me. Item #15 http://www.allbusinessrecords.com/ projects.html.

Table 7.6 Classification task

Worksheet: classification (work in groups)
In the attached page, 15 items are presented
Choose your own criteria and categorize/classify them into sets. An item can belong to several sets. For example, item 1 can belong to set A because it satisfies the "A criterion" and to be included in set B because it also satisfies the "B criterion." In your categorization, focus on the contents or essence of the items and try to avoid "trivial" criteria (e.g., the fact that people appear in items 8, 14, and 15)
Add a new item (not from the given page) to each set
Give a title to each set
Give a title to the whole page

Due to the explanatory power of analogy, when learners face difficulties in understanding a new concept (or an idea), a metaphor may offer a new perspective on the concept and may support its understanding. Yet, although metaphors may help bridge between the new and unfamiliar knowledge (e.g., a variable) and a known knowledge (e.g., a box), metaphors should be used carefully; after all, a metaphor does not refer solely to instances in which the two concepts correspond to each another, but also to contexts in which they do not fit each other. For example, in the case of a variable, the metaphor of a box helps explain the storage property of a variable; at the same time, however, pupils may think that a variable may contain several values (similar to a box that can contain several items). Teachers must be aware of these difficulties and learn how to cope with them (see also Chap. 6).

Activities 53 and 54, to be facilitated in the MTCS course, address metaphors in the context of computer science education.

Activity 53: Metaphors—Preparing a Poster, Variable Exhibition
The students are asked to prepare a poster for an exhibition about the concept of variable. The poster can be prepared with any computational environment that enables to combine text with graphical objects, such as PowerPoint. After the posters are prepared, the students present them in front of the class as an exhibition, and the class discusses pedagogical advantages and pitfalls of each poster. Figure 7.4 presents three illustrative posters which were prepared by prospective computer science teachers for the concept of a variable.

Activity 54: Metaphors—Advantages and Disadvantages of Metaphors
Students are given the worksheet presented in Table 7.7. After they finished their work, a class discussion is facilitated about advantages and disadvantages of metaphors in general and of the specific metaphors they suggested in the worksheet, in particular.

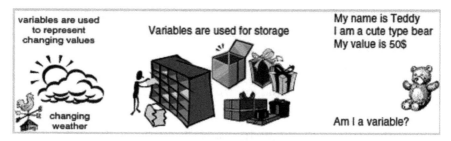

Fig. 7.4 Examples of posters which present metaphors for the concept of variable

Table 7.7 Worksheet about metaphors

Worksheet: Metaphors
For each of the following concepts—a variable, a class, a function
Give at least one metaphor and explain why it is a metaphor
Explain the metaphor limitations
Indicate pedagogical advantages and disadvantages of your metaphors

7.3 Different Forms of Class Organization

Computer science can be taught by lecturing. However, one of the main messages of this guide derived from the constructivist approach, is that this is not the preferable way to enhance learning processes; alternatively, it is argued that in order to learn meaningfully, learners should be active and engaged in the learning process.

In what follows, we first describe several alternative class organizations (to the traditional frontal teaching approach), in which active learning aims to enhance computer science learning. Then, we suggest an activity to be facilitated in the MTCS course to let the students experience these forms of class organization.

Individual work The first, and maybe the simplest, class organization for active learning is when each student works individually on a given task. This class organization is suitable for cases in which a computer science teacher wants to verify that all students are able to cope successfully with a given task or have acquired a specific skill, such as, working in a given Integrated Development Environment (IDE) or practicing a given algorithm.

Working in pairs A class can be organized in pairs, working on either programming tasks or nonprogramming tasks.

In the case of programming tasks, we mention the concept of pair programming (Williams and Kessler 2002), which is one known technique of pair learning in the context of (in most cases) a programming task. In pair programming, a pair of students work on a specific programming task, when one student is the driver, working with the keyboard and the mouse, and the second student is the navigator, who examines the development process from the side, and analyzes, together with the driver, the development process they are going through. The two programmers switch roles frequently; this switch further enhances the learning process of both mates, since in most cases, pair programming guides a problem-solving process that is carried out on two levels of abstraction: the code level and the strategic level.

When this approach is applied to nonprogramming tasks, some advantages of pair programming are not applicable anymore, mainly the code-related aspects. Nevertheless, advantages of pair working can be achieved also when the pair works on a nonprogramming task; it is simply reasonable to assume that in a pair work, the two learners are involved in the problem-solving process, unlike in larger groups, when some students may be more dominant, control the group work, and suppress the involvement and contribution of the other group members.

Working in groups Another form of class organization is working in groups with more than two members. This is a preferable class organization when (a) a computer science teacher observes that a given task requires more than two learners in order to be accomplished, (b) the teacher wishes to exploit team diversity, (c) the class is relatively big and the teacher wants to monitor the groups' work more easily, and finally (d) the class is relatively big, and the teacher wishes to allow all students to be involved and represented in the group presentations presented by one of their team members.

Jigsaw is one technique for class organization in groups, which, according to the Jigsaw Classroom website,[6] carried out as follows. First, the students in the class are divided into groups of five or six students each and their task is to learn a specific topic. Each of the students in each team is responsible for learning a specific part of the topic and to teach this specific part to the whole group. Second, to increase the chances that each report will be accurate, after learning the allocated part of the topic, the students do not return immediately to their jigsaw group, but rather, they meet first with the students from the other groups who have an identical assignment (one student from each jigsaw group) and share what they have learned. In this process, they verify and refine their understanding of the part of the project allocated to them. In the third stage, the jigsaw groups reconvene in their initial heterogeneous configuration and each student of the jigsaw group teaches the other group members the topic he or she has learned which is now considered as her or his specialty. Finally, the teacher can decide whether each group submits a written work, a poster, another group product, or nothing at all.

As can be seen from this short description, the jigsaw classroom organization has many advantages, both cognitive and social, since it enhances learning, listening, cooperation, and knowledge sharing.

Out-of-school Learning To enrich the students' computer science learning experience, out-of-school learning environments, such as museums and exhibitions (see, for example, The Bloomfield Science Museum Jerusalem CAPTCHA exhibition at http://www.mada.org.il/en/exhibitions/captcha), and a tour in a hi-tech company (see, for example, Eidelman et al. 2011), can be integrated in the learning process. When an out-of-school activity is integrated in the learning process, it is important to prepare the students before it takes place and to sum it up after it ends.

---

[6] See http://www.jigsaw.org/.

Distance Learning While up till now we dealt with face to face class organizations, this section deals with distance learning environments, which, due to recent technological developments, are widely accessible now.

A distance learning environment is asynchronic in terms of place and in terms of time—either synchronic or a-synchronic. Though distance learning is not the typical environment for which this guide is dedicated, we found it relevant to include it here since some of its elements can be integrated also in traditional on-campus learning environments. Specifically, we focus here on *MOOC—Massive, On-line, Open Course*—which is one of the newest ways by which distance learning is implemented and represents a new form of class organization (SIGCSE 2014 is the first SIGCSE conference in which MOOCs were discussed explicitly and full sessions were dedicated to this topic; e.g., Warren et al. 2014). This class organization shows that not only computer science is keep developing (as many other sciences), but also new teaching methods and platforms, that use the scientific ideas of computer science, keep developing.

Computer science educators may use a MOOC as learners, when, for example, they should teach a new subject with which they are not familiar, and prefer to learn it in a distance learning environments which provides more flexibility in terms of time and place. In the MTCS course, distance learning can be implemented, for example, for introducing the concept of flipped classroom (see Activity 56).

In this spirit, we decided to include this topic in the book in order a) to expose the prospective computer science teachers to this extremely big and developing educational scene and b) to increase their awareness to its potential integration in traditional on-campus learning environments (alongside its advantages and disadvantages).

The target of Activity 55 is to increase the students' awareness to the different class organizations. It notes that the topic of class organization should not be discussed only in this context; rather, it is important to let the students experience different forms of class organizations throughout the course and to relate to these organizations when appropriate. For example, when Stage A of Activity 52 is facilitated (Sect. 7.2.5), it is relevant to address the power of the group in performing this task.

Activity 55: Different Forms of Class Organizations

Based on students' on-going experience of different class organizations in the MTCS course, an integrative class discussion can address questions such as:

1. What are the differences between the different class organizations?
2. For what purposes each class organization is suitable?
3. What computer science topics are especially suitable to be taught by each class organization?
4. What class organization(s), if at all, is/are especially suitable for computer science learning in general?

Activities 56–57 focus on distance learning and aims to prepare computer science educators to the option of integrating such components in their school or university teaching.

Activity 56: Distance Learning in the School—The Flipped Classroom

One distance learning variation is the *flipped classroom*, where the regular teaching model is flipped: The learning starts with an online session (prepared in advance) learnt by the students individually, usually at home, and only later, in the class, the teacher meets the students for guided work on problem solving, collaboration and interaction.

An optional activity would be to ask students to check one of the flipped classroom websites on any computer science topic and prepare a short presentation to be presented in the classroom. An additional activity would be to ask students to design the second part of the flipped classroom model (the class component) based on a computer science topic taken from one of the flipped classroom websites.

In what follows, a structured activity around this class organization is presented.

- Stage A: Homework toward the lesson on the flipped classroom
  - Find several examples of flipped classrooms and explore their characteristics.
  - Experience one of these examples that deals with computer science education. Document your experience and analyze its pedagogical principles.
  - List pros and cons related to distance learning in general and the flipped classroom in particular.
  - 10-min presentations are a common component in the flipped classroom setting. Prepare a 10-min presentation on any computer science topic according to your choice to be presented for a class that learns in the flipped classroom approach.
    1. List the teaching guidelines you followed.
    2. Reflect on your experience: What came easily? Why? What was difficult? Why?
- Stage B: Teamwork in class
  - Prepare an evaluation rubric for the 10-min presentations prepared at home toward this lesson.
  - Reflect: Are these evaluation criteria differ from evaluation criteria for other teaching formats? Explain.
  - Present your evaluation criteria in front of the class.
  - Upload your evaluation rubric to the forum that was opened for this purpose.
- Stage C: Forum discussion toward the next lesson
  - Assign one of the classmates to moderate a forum discussion.
  - Toward the next lesson: based on the online discussion in the forum, choose one evaluation rubric that is agreed upon by all the students in the class.
- Stage D: Evaluations of the short presentations

Each student presents the 10-min presentation prepared in Stage A; all classmates use the evaluation rubric they agreed upon in Stage C and evaluate the presentation.

- Stage E: Class discussion
  - Each student shares his or her perspective with respect to the evaluation processes.
  - The instructor summarizes the main ideas raised in this discussion.
  - Main pedagogical concepts are highlighted.
- Stage F: Homework, individual work
  - Reflect on the entire process (stages A–E)
  - Suggest improvements for your 10-min presentation

Activity 57: Acquaintance with MOOCs

This activity increases the prospective computer science teachers' attention to the fact that the development process of on-line distance environments is largely based on computer science topic. It is not necessarily suitable for all classes; the instructor should evaluate its suitability to his or her classes.

- Explore what a MOOC is.
- What computer science ideas are relevant when a platform for a MOOC is developed?
- Look at five institutions that offer MOOCs in computer science topics.
  - List these institutions and the computer science topics they teach.
  - What characterizes the offered courses? Address organizational, cognitive, and scientific aspects.

Mobile applications are also new tools that enable new pedagogies in general and forms of class organization in particular. In the spirit of the activity that deals with MOOCs (Activity 57), Activity 58 should be facilitated in classes that may find this topic attractive.

Activity 58: Mobile Applications in Computer Science Teaching

In general, there are two ways by which mobile application can be used for computer science teaching: (a) as a computer game and (b) for the learning of computer science ideas by developing process of applications for mobile devices. (see, e.g., Boticki et al. 2013). In the following questions, refer to these two usages of mobile devices in computer science education.

- In what ways can mobile applications change the teaching and learning of computer science?

- Look for three mobile applications used for computer science teaching. What does characterize them?
- Choose a computer science topic and sketch a mobile application for its teaching. Document the process you went through and reflect on it.

7.4   Mentoring Software Project Development[7]

In what follows, we first address the contribution of project-based learning (PBL) to learners in general and in the context of computer science education in particular. Then, we describe several activities to be facilitated in the MTCS course with respect to the mentoring process of software project development in high school computer science classes.

In PBL situation learners work individually or cooperatively in groups, while the teacher mentors the process of project development. PBL aims to make the learned subject matter relevant for learners and to enable active learning (Blumenfeld et al. 1991). In PBL, learners ask questions, examine their assumptions, design the investigation process, collect and analyze data, use technology and exchange ideas (Krajcik et al. 1999), all while interacting in and with the learning environment in a constructivist fashion (Thomas 2000), and dealing with experiences and deliberations on significant problems (Ernest 1995). Barak et al. (2000) and Waks (1997) assert that PBL makes the learning authentic since it involves learners in activities that are based on daily situations. Further, learners' ownership over the learning process develops their responsibility for their actions, and their cooperative learning with peers involves also social interaction.

Research works indicate that PBL develops thinking practices, independent learner abilities, motivation, self confidence, classmate cooperation, and an integrative understanding of the content as well as of the process (Krajcik et al. 1999; Barak et al. 2000; Green 1998; Shepherd 1998). These results are hardly surprising since PBL enables teachers to adapt the variety of tasks possible in PBL environments to each learner's learning style (Krajcik et al. 1999). At the same time, however, PBL poses some difficulties that learners must face, such as difficulties in coping with the complex and open environment and difficulties with information processes (Krajcik et al. 1998).

In order to help learners deal with a variety of problems, the teacher is required to create an investigation-oriented environment that encourages learners' responsibility and emphasizes an intensive learning process of the project components

---

[7] Based on (Meerbaum–Salant and Hazzan 2008). Copyright 2008 by the Association for the Advancement of Computing in Education (AACE). [http://www.aace.org] Included here by permission.

(Blumenfeld et al. 1991). In general, Waks (1997) asserts that the focus of the teacher's role must be modified in PBL environments from "teaching" activities to "learning" activities, by establishing conditions that enhance learners' curiosity and motivation.

One example of PBL in computer science education is examined in Kay et al.'s (2000) research, which explored different teaching approaches for the Introduction to Computer Science course taught according to the object-oriented approach. PBL was found to be the most suitable approach, since it provides learners with the opportunity to deal with real problem-solving situations and to acquire problem-solving skills and practices. In the same spirit, Johnson (1997) claims that since instructors must teach in a way that develops learners' problem-solving experiences for the benefit of their future work environment, their teaching must guide learners to develop conceptual thinking, criticism, and creativity.

Software development projects conducted within high school computer science classes offer a prime example of a PBL environment. Similar to other PBL situations, the role of the computer science teacher in the development process of a software project is different than his or her role when teaching in the class. In addition to the teaching of the intended programming paradigm, programming language and related computer science concepts, computer science teachers are required to mentor the project development process from the early stages of subject selection to the final stages of testing and verification, to evaluate the learners' learning process and, at the end, to evaluate the developed projects. Fincher and Petre (1998) claim that such a process is a long and complex problem-solving process by itself since it requires computer science teachers to deal with multiple problems simultaneously and to exhibit supervision, management skills of projects of different scales, flexibility, and creative thinking.

This complexity of the teacher's role in mentoring PBL in general and in the context of computer science education in particular, stresses the importance of addressing the mentoring process of software projects in the high school in the MTCS course. Activities 59 and 60 introduce to the students in the MTCS course the potential, as well as the challenges, of mentoring software project development in the high school, and elevate their thinking about how to manage this process in their future classes. Additional activities related to project evaluation are presented in Chap. 10.

Since the mentoring process of software project is not a simple pedagogical task, it is recommended to address this topic in a relatively advanced stage of the MTCS course, after the students have already gained some sense of the essence and spirit of computer science teaching.

Activity 59: Analysis of Mentoring Software Project Development Situations

- Stage A: Watch a video clip
  The following trigger aims to let the students experience, as much as possible, situations they may encounter when guiding pupils in the devel-

opment process of software projects.In this trigger, the students watch a well-selected video clip of a class working on the development of a software project. While the students are watching the video, they are asked to focus on the teacher's behavior, to write down positive and negative characteristics of his or her behavior and to imagine how they would act in similar situations. The video can be paused from time to time for short discussions.

It is important to select the video very carefully so that it indeed presents different kinds of situations and different teaching approaches. If it is selected properly, such a trigger and the stages that follow it let the students experience some of the complexity involved in guiding pupils in the development of software projects.

As mentioned in Sect. 8.2, if such a video is not available, it is possible to visit a real high school computer science class whose pupils develop software projects; alternatively, if this is not an applicable option, the instructor of the MTCS course can start from Stage C of this activity.

- Stage B: Class discussion

  After the video is watched, a class discussion takes place, in which the behavior of the computer science teacher is analyzed. Here are several questions that the discussion can concentrate on:

  1. What should a computer science teacher know for mentoring software project development?
  2. What challenges does a computer science teacher face in this process?
  3. What are the pedagogical advantages of software project development by computer science learners? What are the disadvantages of such situations?

- Stage C: Worksheet on PBL situations in computer science education, group work

  The students work in groups on the worksheet presented in Table 7.8, whose purpose is to concentrate the students' attention on the details of class management situations in a computer lab in which learners develop software projects.

- Stage D: Class discussion

  The dialogues developed by the students in Stage C are presented in front of the class and discussed. It is reasonable to assume that different kind of responses will be suggested for each statement. In all cases, however, it is appropriate to analyze what aspects are addressed in the presented scenarios: Did it focus on computer science concepts? Was the help given by the computer science teacher technical? Did the teacher's response address pupil's emotions? Did the teacher's response address other motivational factors?

Table 7.8 Worksheet on PBL situations in computer science

Worksheet
The following statements were said by high school pupils while working on their software projects in the computer lab
Assume that you are the computer science teacher of this class
Select five statements and for each of them:
Describe your reaction to the pupil's statement
Explain why you decided to answer the pupil in this particular way
Speculate the pupil's reaction to your response and the continuation of the dialogue between you and the pupil
Pupils' statements
I do not know how to start
Why did I choose this project? I am so stupid
I cannot do it
How does the function XXX work?
I need a function that does….
I am so satisfied with my progress
The program does not do what I wanted it to do
The computer did not save the last version of my project. I quit!
I am raising my hand for half an hour and you do not approach me
How can I start working on it?
I do not want to present my project in front of the class
I do not understand the computer's response. Why did it print this message?
It does not work
How can I get 100 in the project?
What should I do now?

- Stage E: Summary
  The summary of this activity should address the following issues, and, if relevant, additional topics:
  - The mentoring of software project development in the high school is a complex task both from the technical perspective and the emotional perspective.
  - The mentoring of software projects requires the computer science teacher to respond simultaneously to many pupils.
  - When a computer science teacher responds to pupil's question, the teacher should not solve the problem for the pupil, but rather guide the pupil in a way that enables the pupil to move on.

Activity 60: Scheduling the Mentoring Process of Software Project Development

- Stage A: Setting the framework
  The instructor, together with the students, sketches a pedagogical environment in which a computer science teacher mentors a class of 20 pupils in the development process of a software project. This sketch includes: the development environment, the programming paradigm and programming language, the project scope, the length of the development period, and the grading policy.
- Stage B: Group work
  The students are asked to build a schedule for the mentoring process of the class they just sketched. For each period of time of the entire development process, they are asked to indicate the main activities that the computer science teacher and the pupils accomplish with respect to the project development. They should also explain each of their pedagogical considerations. This process can be repeated with respect to different periods of time, for example, 1 school year, 3 months, etc.
- Stage C: Class discussion
  Following the group work, the different options proposed by the groups are presented, together with their pedagogical considerations. In this discussion, it is important to address several questions, such as:
  - Should a computer science teacher teach first all the relevant material and only then start guiding his or her pupils in the development process? Or, alternatively, should the computer science teacher integrate the project development process with the actual teaching of the computer science material? What advantages and disadvantages does each approach have?
  - In what ways are the different-in-length schedules that the students suggested similar? In what ways are they different?
  - What factors did the students consider when building these schedules: the development environment? The programming paradigm? The programming language? Other factors? In what way?
  - What aspects, beyond the actual teaching of computer science content, should a computer science teacher pay attention to when mentoring software project development?
- Stage D: Read a paper, homework
  If the instructor of the MTCS course wishes to further deepen the students' attention to managerial aspects of mentoring software projects in high school computer science classes, the students can be asked to work on the homework presented in Table 7.9.

Table 7.9 Homework about mentoring methodology for software projects

Homework—Mentoring methodology for software projects
The following paper presents a mentoring methodology for high school computer science pupils who develop a software project
Meerbaum–Salant, O. and Hazzan, O. (2010). An agile constructionist mentoring methodology for software projects in the high school, ACM Transactions on Computing Education—TOCE 9(4)
Read the paper and analyze the mentoring methodology it presents Address the following questions: What are the advantages of the mentoring methodology? What are its disadvantages? Can you suggest improvements to the mentoring methodology presented in the paper?

References

Barak M, Waks S, Doppelt Y (2000) Majoring in technology studies at high school and fostering learning. Learn Environ Res An Int J 3:135–158

Blumenfeld PC, Soloway E, Marx R et al (1991) Motivating project-based learning: sustaining the doing, supporting the learning. Educ Psychol 26:369–398

Boticki I, Barisic A, Martin S, Drljevic N (2013) Teaching and learning computer science sorting algorithms with mobile devices: a case study. Comput Appl Eng Educ 21:E41–E50. doi:10.1002/cae.21561. http://onlinelibrary.wiley.com/doi/10.1002/cae.21561/abstract

Eidelman L, Hazzan O, Lapidot T, Matias Y, Raijman D, Segalov M (2011) Can a 2-hour visit to a hi-tech company increase interest in and change perceptions about computer science? ACM Inroads mag 2(3):64–70

Ernest P (1995) The one and the many. In: Steffe LP, Gale J (Eds) Constructivism in education. Lawrence Erlbaum Associates, Hillsdale, pp 459–486

Fincher S, Petre M (1998) Project-based learning practices in computer science education. Proc Front Educ Conf, Tempe Arizona:453–494

Green AM (1998) Project-based learning: Moving students through the GED with meaningful learning. ERIC Database, ED422466

Hebetim (1995) Educational game—the Conditional-Statement-Bingo, Hebetim—Journal of the Israeli National Center for Computer Science Teachers, June:31–32

Johnson DS (1997) Learning technological concepts and developing intellectual skills. Int J Technol Des Educ 7:161–180

Kay J, Barg M, Fekete A et al (2000) Problem-based learning for foundation Computer Science courses. Comput Sci Educ 10:109–128

Krajcik JS, Blumenfeld PC, Marx RW et al (1998) Inquiry in project-based science classrooms: Initial attempts by middle school students. J Learn Sci 7:313–350

Krajcik JS, Czerniak C, Berger C (1999) Teaching science: A project- based approach. McGraw Hill College, New York

Lakanen A, Isomöttönen V, Lappalainen V (2014) Five years of game programming outreach: Understanding student differences. Proceedings of SIGCSE 2014—The 45th ACM Technical Sym on Comp Science Edu, Atlanta, GA, USA: 647–652

Lakoff G, Johnson M (1980) Metaphors we live by. The University of Chicago Press, Chicago IL

Lapidot T, Levy D (1993) From programming to computer science: Opportunities and pitfalls. In: Kynigos C (ed) Proceedings of the 4th Euro-Par conference. Logo conf, Athens

Levy D, Lapidot T (1997) Rich task: Opportunities for learning computer science ideas. Hebetim—J Israeli Natl Cent for Comput Sci Teach 9:34–26

Meerbaum-Salant O, Hazzan O (2008) Challenges in mentoring software development projects in the high school: analysis according to Shulman's teacher knowledge base model. J Comput Math Sci Teach 28(1):23–43

Meerbaum–Salant O, Hazzan O (2010) An agile constructionist mentoring methodology for software projects in the high school, ACM Transactions on Computing Education—TOCE 9(4), Article 21:1–29

Novak J D, Cañas A J (2008) The theory underlying concept maps and how to construct them, Technical Report IHMC CmapTools 2006 – 01 Rev 01–2008, Florida Inst. for Hum. and Mach. Cogn. http://cmap.ihmc.us/Publications/ResearchPapers/TheoryUnderlyingConceptMaps.pdf. Accessed 14 July 2010

Shepherd H G (1998) The probe method: a project-based-learning model's effect on critical thinking skills. Diss Abstr Int Sect A 59(3 A):779

Thomas J W (2000) A review of research on project-based learning. http://www.autodesk.com/foundation.

Waks S (1997) Education and technology-dimensions and implications. Position paper on prospects of interrelationship between the academia and the educational system in Israel, the Van Leer Jerusalem Institute, The Forum for High. Educ., the Ministry of Education, Culture and Sports, Israel

Warren J, Rixner S, Greiner J, Wong S (2014) Facilitating human interaction in an online programming course. Proceedings of SIGCSE 2014—The 45th ACM Technical Sym on Comp Science Edu, Atlanta, GA, USA, 665–670

Williams L, Kessler R (2002) Pair programming illuminated. Addison Wesley, Boston

Lab-Based Teaching

8

Abstract

This chapter focuses on computer science teaching methods that fit especially to be employed in the computer lab. The uniqueness of the computer lab as a learning environment for computer science is explained by the fact that it enables learners to explore their problem-solving strategies, to express their solutions to a given problem, to get feedback regarding to the correctness of their solution and to reflect on it, to develop large projects, to explore new topics, and to deepen their understanding of the nature of the algorithms they develop. The main purpose of the lessons in the Methods of Teaching Computer Science course is to expose the students to usages of the computer lab as a learning environment and to let them realize how it may improve their future pupils' understanding of computer science ideas. One of the main messages of this chapter is that the learning of computer science in the computer lab is not limited to programming tasks; rather, the computer lab can be used in additional pedagogical ways that further enhance learners' understanding of computer science. Specifically, the following topics are addressed in this chapter: what is a computer lab?, the lab-first teaching approach, visualization and animation, and using the Internet in the teaching of computer science.

8.1 Introduction

Clearly, the learning of computer science should take place, at least partially, in the computer lab. This learning environment enables learners to explore their problem-solving strategies, to express their solutions to a given problem, and to deepen their understanding of the nature of algorithms they program. Further, the learning of computer science in the lab has the potential to increase learners' understanding of the essence of computer science—what can be done with computers, that is, what is computable, as well as the influence of the field of computer science on the world.

© Springer-Verlag London Limited 2014
O. Hazzan et al., *Guide to Teaching Computer Science*,
DOI 10.1007/978-1-4471-6630-6_8

Lately (effective from fall 2014), the College Board (*apcentral.collegeboard.org*) has decided to reinforce the lab component of the AP Computer Science A course with "a substantial laboratory component in which students design solutions to problems, express their solutions precisely, test their solutions, identify and correct errors (when mistakes occur), and compare possible solutions." The course description emphasizes that although the "course draws heavily upon theory, formal logic, abstract data structures, and a conceptual understanding of algorithms, students also must gain significant experience applying the concepts to tackle a wide range of problems. As students design data structures and develop algorithms, they also integrate ideas, test hypotheses, and explore alternative approaches." Further, it is also stressed that "activities motivated by real-world applications can provide insights about how computing can be useful in society, motivate the study of technical issues, and capture students' interest."

This chapter focuses on computer science teaching methods that are suitable, especially, to be employed in the computer lab, and specifically, on the lessons in the Methods of Teaching Computer Science (MTCS) course that aim to expose the students to pedagogical usages of the computer lab that target to improve their future pupils' understanding of computer science ideas.

One of the main goals of the ideas presented in this chapter is to let the students realize that the learning of computer science in the computer lab is not limited to programming tasks. Rather, they, as future computer science teachers, can use the computer lab in additional ways that further enhance learners' understanding of computer science. We mention, though, that the usage of the computer lab for programming tasks is meaningful in order to provide learners with opportunities to gain some programming experience. In this spirit, the lessons in the MTCS course, which are dedicated to lab-based learning, highlight the added value of the computer lab for learners' understanding of computer science concepts, beyond the advancement of their programming skill. This added value includes benefits of project-based learning (Sect. 8.7.4) and the conception of the computer lab as a learning environment in which learners experiment and check hypothesis as recommended by the College Board (2014). This assertion is based on the constructivist approach presented in Chap. 2.

This chapter first elaborates on the computer lab as a learning environment and then presents several lessons that can be facilitated in the MTCS course which examine, from different perspectives, the role of the computer lab in computer science teaching and learning processes. Specifically, the following topics are addressed:

- What is a computer lab? (Sect. 8.8.2)
- The lab-first teaching approach (Sect. 8.8.3)
- Visualization and animation (Sect. 8.8.4)
- Using the Internet in the teaching of computer science (Sect. 8.8.5)

Clearly, this is not an exhausting list and additional pedagogical ways exist that fit to be applied in the computer lab for computer science teaching; they do, however, represent a variety of usages, from which each instructor of the MTCS course can choose to address the ones that fit him or her own pedagogical approach.

8.2   What Is a Computer Lab?

A laboratory (lab) is a common concept in any science. For example, according to the *Fourth Edition of The American Heritage® Dictionary of the English Language*,[1] a laboratory is

1. a. A room or building equipped for scientific experimentation or research
   b. An academic period devoted to work or study in such a place

2. A place for practice, observation, or testing

Another definition for a lab (similar in some senses to the first one) is "A place equipped for experimental study in a science or for testing and analysis" (*Merriam-Webster's Medical Dictionary*, © 2002 Merriam-Webster).

As can be seen, these definitions emphasize the experimental aspect of the lab, when experiment is defined by Fourth Edition *of The American Heritage® Dictionary of the English Language* as

A test under controlled conditions that is made to demonstrate a known truth, examine the validity of a hypothesis, or determine the efficacy of something previously untried.

and the corresponding verb—experiment—is defined as follows:

1. To conduct an experiment
2. To try something new, especially in order to gain experience: *experiment with new methods of teaching*

The importance attributed to the lab is not limited to scientific *research* and is expressed also in the context of science *teaching*. The educational literature is full with praises on the advantages and contributions of laboratory work to the learning process. For example, Nersessian (1991) claims that "hands-on experience is at the heart of science learning." According to Ma and Nickerson (2006), "there is no doubt that lab-based courses play an important role in scientific education" (p. 2).

Specifically, in sciences, such as biology, chemistry, and physics, experiments that learners perform in the lab aim, in many cases, to let the learners be active, rather than passive observers of the scientific world. This pedagogical target is achieved by the design of experiments that demonstrate and illustrate to the learners what is taught theoretically in the class (either before or after the lab experience), guide the learners to check hypothesis, train them how to perform experiments, let them practice data collection and analysis methods, teach them research skills, and foster their critical thinking.

In the context of computer science education, the lab concept is captured similarly but with several slight differences. First, the physical structure of the lab in the two cases—the lab in science teaching, such as biology and chemistry, and the lab in the case of computer science teaching—is different. Specifically, these two labs

---

[1] Copyright © 2000 by Houghton Mifflin Company, Published by Houghton Mifflin Company.

are different in terms of their equipment and its usage; while in the above sciences the equipment includes test tubes, materials, and other physical instruments that learners use to perform experiments, the image of the lab in the context of computer science education is a computer lab—a room with computers in which the learners work, mostly—program.

According to Petre (2011) "An informal scan through a variety of papers about 'experimentation' in CS education suggests that 'experimentation' in our discipline more often means 'building and trying things out' than 'observing, hypothesizing and testing systematically'." Petre argues that we need both kinds of experimentation in computer science and bases the importance attributed to the second perspective on "Years of observation and interview of expert software developers in industry [which] reveal that scientific enquiry and dialogue is part of the professional repertoire that contributes to their innovation and success."

This perspective at the computer lab as a place, in which experiments are carried out, further highlights the scientific aspect of computer science (see Chap. 3). Indeed, as a science, computer science has its own scientific exploratory methods which include, as in other sciences, the experimental path that starts with hypothesis, continues with experiment and data gathering, and ends with the data analysis and conclusion formulation. We suggest that it is important to deliver this message to the prospective computer science teachers.

In Sect. 8.8.3, about the lab-first approach, we demonstrate how the computer lab serves as a place for carrying out experiments and checking hypotheses.

The lab in the context of computer science teaching and learning has several additional advantages:

- Learners are familiar with how to work with the equipment in this lab, that is, how to work with computers, so there is no need to teach them how to use this equipment.
- The work in the computer lab does not require the preparation of any physical material, so a computer science teacher can be flexible, and when observing that it is a suitable time to explore a specific topic with the computers, he or she can just ask learners to use the computers (if the lesson takes place in the lab).
- Since no physical material is required in the computer lab, budget constrains should not limit or block the use of the lab for computer science learning purposes.
- In the computer lab, an experiment can be run many times, one after the other, with an immediate feedback provided to learners by the computer.

While not claiming that all computer science courses should be lab based, most courses can benefit from the use of some well-designed laboratory components (e.g., Knox et al. 1996; Walker 2011; Lee 2013). In these lessons, the teacher's role is very important and required considerable preparation. The teacher can demonstrate an experiment to the pupils or, alternatively, guide them, in an active learning manner, to explore a new topic. As has been asserted before in this guide, the more learners are active, their learning is more meaningful, and, in the context of lab learning, the benefits of the lab as a learning environment are exploited more efficiently. It should be remembered, though, that the exact doze of lab-based teaching should be seriously considered in each case in order to vary the teaching methods employed in the said teaching situation. This important role of the teacher in lab-based teaching situations explains the importance of including this topic in the MTCS course.

Activities 61 and 62 illuminate the computer lab as a learning environment. During their facilitation, central themes of the computer lab, mentioned above, can be integrated.

Activity 61: Analyzing a Computer Science Lesson in the Computer Lab

The students are presented with a video clip that takes place in the computer lab. After the clip is presented, the instructor of the MTCS course and the students analyze it, addressing questions such as: What is unique in this situation? In what ways does it differ from the traditional way of teaching computer science? What is the teacher's role in this situation? The role of this discussion is to identify the advantages and challenges of learning and teaching computer science in the computer lab (as described above). The advantages and the challenges involved in this learning environment are addressed also in the continuation of this chapter, with respect to the different usages of the computer lab in computer science education.

If such a clip is not available, a similar activity can be facilitated based on students' observation of a real lesson that takes place in a high school computer lab. Prior to the actual observation, the students should be guided by questions, such as: How does the teacher behave in the lesson? How are the lessons managed? How do pupils behave? How do the pupils behave when they face difficulties? How are tasks presented to the pupils? Do pupils work individually, in pairs, in groups? If they work in groups, how is the work divided among the pupils in the group?

Activity 62: A "Dry" Lab

This activity is inspired by *The Little LISPer* (Friedman and Felleisen 1986) book, which demonstrates a pedagogical approach which is closely related to the constructivist approach present in Chap. 2. According to this approach, learners can gain better understanding of the concepts they learn by forming their own definitions based on a guided exploration, than by being presented with well-defined terminology of the learned concepts by the teacher.

The students get the worksheet presented in Table 8.1, in which they are asked to reveal the meaning of the instructions of a programming language called DL (dry lab). The worksheet is worked on without computers. In computer science classes, the worksheet can be worked on with respect to general programming languages.

Table 8.1 "Dry" lab worksheet

Worksheet: The DL language

The following is a list of instructions in the DL programming language together with their output

The instruction	The output
DL.write [2014]	2014
DL.write [# laboratory #]	Laboratory
DL.write [# the rules of #! # DL are #! # easy #]	The rules of DL are easy
DL.write [5*10! # shalom #]	50 shalom
DL.write [% x=13%! # x=#! x]	x=13
DL.write [% x=13%! # x=54 #]	x=54
DL.write [% x=9%! x! % x=x+1%! x*2]	9 20
DL.write [# x=#! x]	x=what?
DL.write [y]	What?
DL.write [# x=9 #! x! # print end #]	x=9 what?

Complete the following:

a. The output of an instruction of the pattern DL.write [number] is _____

b. The output of an instruction of the pattern DL.write [# string #] is _____

c. The output of an instruction of the pattern DL.write [arithmetic expression] is

_____

d. The role of % is _____

e. The role of ! is _____

f. The message *what?* is written when _____

g. To get the output "DL.write is what?", the following instruction(s) should be executed: _____

h. To get the output: "what? = what?", the following instruction(s) should be executed: _____

After the worksheet is worked on by the students, it is important to facilitate a reflective session in which, in addition to the students' exploration strategies, the question whether this activity is a lab activity is examined. Another question that can be asked is what characterizes a good lab-based lesson.

In what follows we present a collection of statements, offered by a group of in-service high school computer science teachers after they had worked on the above worksheet, which reflects their associations with what a lab-based teaching is in the context of computer science. As can be observed from these statements, this activity can be considered as a fruitful trigger for the discussion about lab-based teaching.

- Is the above activity a lab activity?
  - Yes, it is a lab because it involves query, experience, activity, learning, knowledge construction.
  - It is not a lab because we did not use computers; learners do not work with any mediator and therefore it is impossible to really check [hypothesis].
  - It is a lab since a lab is everything that is based on active learning and not on teacher presentation, and it also includes teacher–students discourse.
  - A lab is a place to check hypothesis and something should be provided to the learners to check whether they understood correctly; therefore, this worksheet is not a lab-based activity.
- What is a good lab-based lesson?
  - A good lab integrates colors, shapes, sounds, and motion.
  - If the lab-based lesson is implemented well, the teacher is unemployed.
  - In a good lab-based lesson, students succeed learning computer science ideas.
  - In a good lab-based lesson, pupils teach pupils.
  - In a good lab-based lesson, pupils enjoy, are busy, concentrated, and have high spirits.
  - It is a good lab-based lesson, when a pupil says "now I got it."
  - It is a good lab-based lesson, when a weak pupil has bright eyes.
  - In a good lab-based lesson, excellent pupils are well challenged.
  - In a good lab-based lesson, weak pupils are well challenged.
  - In a good lab-based lesson, pupils invent solutions.

As can be observed, even though this activity focuses on a "dry" lab, it increases the teachers' awareness to key terms of lab-based learning, such as, inquiry, experiment, checking hypothesis, learning, and knowledge construction.

8.3 Lab-First Approach

As mentioned, the computer lab can be viewed as a place in which experiments are carried out, hypotheses are checked, and conclusions are derived following an experimental process. In this context, activities, such as checking program correctness and program behavior with different inputs, are carried out in the computer lab on a regular basis.

In this section, we further pursue this perspective and examine the lab-first pedagogical approach which implements the constructivist approach and demonstrates an active-based learning teaching strategy (see Chap. 2). According to the lab-first approach, learners explore first a computer science topic in the computer lab, and after they gain some experience with the said topic, and based on this experience, they are introduced to the said concept in the class. As can be seen, the lab-first approach reverses the traditional teaching order, by which a theoretical material is presented first, following by its practicing and experimenting in the computer lab[2].

The lab-first approach has both advantages and disadvantages. On the one hand, in the spirit of constructivism, its main advantage is expressed by the active experience learners get in the computer lab, which in turn, establishes foundations based on which learners construct their mental image of the said topic; on the other hand, the lab-first teaching approach involves some insecurity feelings expressed both by the computer science teacher and the learners.

One way to deliver this complex message to the students in the MTCS course is by letting them experience the lab-first approach. Activity 63 was designed for this purpose by guiding the students to explore the lab-first teaching approach both from the learners' and the computer science teacher's perspectives. The worksheet presented in this activity is only one option by which the lab-first approach can be implemented in general and by which the conditional statement *if* can be introduced to novice learners, in particular. For example, the worksheet can include more open tasks, such as the following one: After Task 3, an intermediate summary of the general structure of the *if* statement is presented. An open assignment would ask learners to explore the extent of *instruction for execution* by themselves. In this spirit, the worksheet would present the following instruction: So far, the only instruction used inside the *if-statement* was *System.out.println*. Replace the *System.out.println* with different java instructions and check which instructions can be nested in the *if-statement*.

If the instructor of the MTCS course wishes to let the students feel the learning experience similar to their future high school pupils, it is recommended to use the lab-first approach with one of the integrated development environments (IDEs) presented in Activity 67 (see Sect. 8.8.4) for which they are not familiar.

Activity 63: Pedagogical Exploration of the Lab-First Teaching Approach

- Stage A: First experience with the lab-first teaching approach, work in pairs on the computer
  The students are given the worksheet presented in Table 8.2 (Paz 2006). They are told that high school computer science pupils had worked on this worksheet before they learned the topic of conditional statements. The students in the MTCS course are asked to:

1. Work on the activities presented in the worksheet.
2. Analyze the worksheet from a pedagogical perspective: What pedagogical advantages does it have for the introduction of a new topic? What is the purpose of each task presented in the worksheet? How does each task support learners' understanding of conditional statements?

---

[2] In this sense, the lab-first approach is similar to the flipped classroom (see activity 56, section 7.3).

Table 8.2 Lab-first worksheet (This worksheet was developed by Dr. Tamar Paz and is included here with her permission. See Paz (2006, pp. 33–36))

Worksheet

The Conditional Statement If
Conditional statements enable to write computer programs that act in one way if a given condition is fulfilled, and to act differently if the condition is not fulfilled.

```
import java.util.Scanner;
public class Condition {
  public static Scanner input = new
  Scanner(System.in);
  public static void main(String[] args) {
    double num;
    System.out.println("enter number");
    num = input.nextDouble();
    System.out.println("good number");
    System.out.println("end");
  }
}
```

Task 1, Part A
Open a new class and type the class presented on the left.
Save it, run it, check its output and write it down.

```
import java.util.Scanner;
public class Condition {
  public static Scanner input = new
  Scanner(System.in);
  public static void main(String[] args) {
    double num;
    System.out.println("enter number");
    num = input.nextDouble();
    if (num > 0)
      System.out.println("good number");
    System.out.println("end");
  }
}
```

Task 1, Part B
We now change the class so that the message "good number" is printed only if the variable num is positive.
Change the class according to the code on the left. Save it and run it.
When the program waits for an input, type a positive number.
The output is: _____
Run it again. This time, when the program waits for an input, type a negative number.
The output is: _____

Complete the following sentence:

The meaning of the statement
    if (num > 0)
        System.out.println("good number");
is: if _____
    then _____

The general structure of the conditional statement *if* is:
    if (a condition to be checked)
        a statement for execution;

Task 2, Part A

Write a program that prints "Good" for a given grade if it is higher than 80.
Save it and run it several times. Each time type a different grade and verify that the message is printed only when the grade is higher than 80. Try also the number 80.

Task 2, Part B

Add to your program a conditional statement so that if the grade is lower than 55, the program prints the message "Try Again".
Save and run it several times. Each time type a different grade and check that the correct message is printed.

```
import java.util.Scanner;
public class License {
  public static Scanner input = new
  Scanner(System.in);
  public static void main(String[] args) {
    double age;
    System.out.println("enter age");
    age = input.nextDouble();
    if (age>=17)
```

Task 3, Part A
The class on the left includes an operation that declares whether a driver license can be obtained for a given age.
Type the class.
Save it and run it three times. Each time, type a different age and fill in the following table:

Table 8.2 (continued)

```
      System.out.println("license");
   }
}
```

```
import java.util.Scanner;
public class License {
   public static Scanner input = new
   Scanner(System.in);
   public static void main(String[] args) {
      double age;
      System.out.println("enter age");
      age = input.nextDouble();
      if (age>=17)
         System.out.println("license");
      else
         System.out.println("no license");
   }
}
```

The typed number			
Is "license" printed?			

Task 3, Part B
We now expand the conditional statement so that if the age is not bigger or equal 17, "no license" is printed.
Change it and save it.
Run it three times. Each time, type a different input and fill in the following table:

The typed number			
The printed message			

Intermediate summary:
The statement
```
   if (age> = 17)
      System.out.println("license");
   else
      System.out.println("no license");
```
is an extended conditional statement.
Its general structure is:
```
   if (a condition)
      an instruction for execution;
   else
      an instruction for execution;
```

Task 4
Write a program that prints the bigger number from two given numbers.
Type it, save it, and check that you get the expected output.
```
import java.util.Scanner;
public class Check5 {
   public static Scanner input = new
   Scanner(System.in);
   public static void main(String[] args) {
      double x, y;
      System.out.println("enter two numbers");
      x = input.nextDouble();
      y = input.nextDouble();
      if (x < y)
         System.out.println("hello");
   }
}
```

Task 5
Type the given class.
The left column of the table below includes different conditions.
Run the class 6 times. Each time, change the condition (x < y) with one of the conditions presented in the table, and fill in the table.

The condition	The sign		Its meaning
x < y	<		Smaller than
x <= y	<=		
x > y	>		
x >= y	>=		Bigger or equal

Table 8.2 (continued)

x ! = y	!=
x == y	== (Note: These are two equal signs)

Task 6, Part A
Write a program that prints whether a given number is even or odd.
Hint: Use the operation %.
Type, save, and run the program. Check that its works properly.

Task 6, Part B
Write a program that prints whether the first given number is divided by the second given number without a
    reminder.
Type, save, and run the program. Check that its works properly.

Task 7
Write a program that prints whether February has 28 or 29 days in a given year.
If a year number is divided by 4 (e.g., 2008), February has 28 days; otherwise, February has 29 days.
Type, save, and run the program. Check that its works properly.

Task 8
According to the rules of the Youth Movement, in the summer camp, each coach should supervise 10
    members. Accordingly, if the number of members is 10 or less, 1 coach is sufficient; if the number of
    the members is 11–20, two coaches should be allocated, etc.
Write a program that prints the number of coaches that should be assigned to supervise the given number of
    members who enrolled the summer camp.
Hint: Use the operations / and %.
Type, save, and run the program. Check that its works properly.

We note that in order to let the students in the MTCS course taste this teaching method, it is sufficient to let them work only on the beginning of the worksheet.

The continuation of this worksheet on conditional statements teaches the following topics: blocks, logical connectors, the Boolean data type, and the conditional statement switch.

- Stage B: Class discussion
  After the students worked on the worksheet presented in Table 8.2, a class discussion takes place in which their pedagogical comments are addressed. In this discussion, it is important to address guidelines that computer science teachers can follow when they design worksheets based on the lab-first teaching approach. Here are several such guidelines:
- The topic (or aspect) on which the lab-first activity focuses was not taught yet in the class.
- Active learning should be fostered.
- The topic should be divided into subtopics.
- The concepts addressed in the lab should be identified and listed, as well as connections between them.
- Each task should have a well-defined purpose and should concentrate on one specific idea.

- Intermediate conclusions that summarize what has been learned so far should be included in the lab-first worksheet. These conclusions should not necessarily include formal definitions; however, they should be intuitive enough to allow learners go on working on the worksheet. The exact formal terminology can be presented in the lesson that takes place in the class after the lab.
- Since the worksheet guides learners to formulate their conclusions based on what they just observed, each time only one specific conclusion should be addressed.
- The worksheet should include enough space to enable learners write their answers on the worksheet. Learners are distracted when asked to read instructions from a worksheet and to write their answers in their notebook.
- After a concept is introduced in the worksheet, learners should work on several tasks that employ this concept in order to let them practice what they have just learned.
- Since learners have different learning rhythms, it is important to provide advanced learners with advanced tasks to work on when other learners still study the basic material. This can be done by including at the end of the worksheet challenging (advanced) tasks that not all the learners should solve.
- Checking crossroads should be included in a lab-first worksheet. A checking crossroad is an assignment that checks learners' understanding of a specific concept/topic(s) before they move on studying the next one.
- Following Petre (2011), it is important to design the lab "thoughtfully with sufficient degrees of freedom to enable discovery and accommodate a variety of studies, but not so much freedom that students become distracted or confused" (p. 19).
- Stage C: Development and analysis of a lab-first worksheet, homework
  The students are asked to select a computer science topic and to develop a worksheet that teaches the selected topic in the lab-first teaching approach. For each question/task presented in the worksheet, the students should explain its purpose and the pedagogical principles that guided the task formulation.

Another topic that can be discussed in the MTCS at this stage is the kinds of questions that fit to be presented to learners in the lab-first teaching approach. Here are several kinds of such tasks: analysis of the behavior of a given program, filling in specific parts in a given program, or writing a computer program that fulfills specific requirements. See also Chap. 9 for additional types of questions.

If time allows, additional questions can be addressed and discussed with the students in the MTCS course: Is it always possible to apply the lab-first teaching approach? Does this approach fit all learners, all teachers, specific types of learners, specific types of teachers? What class organizations fit for the lab-first approach (see Sect. 8.7.3)?

We summarize several attributes of the lab-first teaching approach:

- The lab is the first step in learners' learning of a new topic (such as the conditional statement *if*) or aspect (such as efficiency or abstraction). After the lab, a lesson should take place in the class, in which the computer science teacher, together with the learners, summarizes what was learned in the lab.
- Although the lab-first teaching approach has many pedagogical and cognitive advantages, some teachers avoid using it from different reasons, mainly because it changes the traditional teaching approach and the design of such worksheets requires the expression of creative thinking.
- When the lab-first teaching approach is applied in the first time, it is new and unfamiliar both to the computer science teacher and to the pupils. This fact should be taken into the considerations and immediate conclusions about its usefulness should be avoided. After it is experienced several times, however, both the computer science teacher and the pupils may start realizing its advantages and the challenges it introduces.

8.4 Visualization and Animation

According to Wikipedia,[3] "the objectives of software visualizations are to support the understanding of software systems (i.e., its structure) and algorithms (e.g., by animating the behavior of sorting algorithms) as well as the analysis of software systems and their anomalies (e.g., by showing classes with high coupling)." Price et al. (1998) distinguish between program and algorithm visualization. Based on Price et al. (1998), Urquiza-Fuentes and Velázquez-Iturbide (1998) give the following definitions for these terms:

Algorithm Visualizations: The static or dynamic visualization of higher-level abstractions which describe the software.

Program Visualizations: The visualization of actual program code or data structures—low-level abstraction—in either static or dynamic form.

Naps et al. (2003) claim that "the impetus for visualization in computing comes from the inherent abstractness of the basic building blocks of the field. Intuition suggests that, by making these building blocks more concrete, graphical representations would help one to better understand how they work." However, based on Naps et al. (2003), Rössling and Velázquez-Iturbide (2009) claim that though "intuitively, most educators agree about the great potential of software visualization for computer science education, adoption of software visualization is lower than developers of program or algorithm visualization systems would expect." They summarize a survey, conducted among educators by the ITiCSE 2002 Working Group on "Improving the Educational Impact of Algorithm Visualization" (Naps et al. 2003), which determined the most frequently cited factors that made educators reluctant or unable to use software visualizations. Among these factors they mentioned:

---

[3] Source: http://en.wikipedia.org/wiki/Software visualization.

- Lack of time for different tasks (e.g., to learn a new tool or to develop a visualization).
- Technical issues of software visualization tools (e.g., lack of effective development tools or lack of reliable software).
- Integration into the courses or the classroom (e.g., time to adapt visualizations to a course or visualizations may hide important details or concepts).
- Other factors (including lack of evidence of educational effectiveness).

Ben-Bassat Levy and Ben-Ari (2007) attribute two primary causes to teachers' negative experience with animation systems: "First, a pedagogical software tool cannot stand on its own; rather, it must be integrated into the curriculum through other learning materials such as the textbook. […] Second, to the extent that a software tool is intended for independent use by students as opposed to demonstrations during frontal instruction by the teacher, the issue of the centrality of the teacher must be taken into account. Centrality appears to be an issue both for experienced and highly confident teachers, and for those with little experience and low self confidence." (p. 250). They conclude that their research highlights "the extreme importance of issues relating to control. It is not enough to develop a beautiful and pedagogical useful tool; issues such as easy installation, training courses, and tutorials are of equal importance because they will increase an educator's feeling of control. Similarly, training courses should not ignore operational or pedagogical difficulties that can arise from the use of a software tool. They should address the changing role of the educator when using the tool, emphasizing that they remain in control and do not relinquish their central position" (Ben-Bassat Levy and Ben-Ari (2008, p. 172).

Shaffer et al. (2010) present findings regarding the state of the field of algorithm visualization based on analysis of a collection of over 500 algorithm visualizations. They state that many algorithm visualizations are of low quality, and coverage is skewed toward a few easier topics and suggest that this can make it hard for instructors to locate what they need.

In order to deepen the understanding of how learners can be involved in an educational environment that includes visualization, Naps et al. (2003) define an Engagement Taxonomy. This taxonomy is based on six different forms of learner engagement with visualization technology, as is described very briefly in what follows:

- **No viewing:** No visualization technology is used at all.
- **Viewing:** Viewing by itself is the most passive form of engagement, but at the same time is the core form of engagement, since all other forms of engagement with visualization entail some kind of viewing.
- **Responding:** The key activity in this category is answering questions concerning the visualization presented by the system.
- **Changing:** The key activity in this category allows learners to change the input of the algorithm under study in order to explore the algorithm's behavior in different cases.

- **Constructing:** Learners construct their own visualizations of the topic under study.
- **Presenting:** A visualization is presented to an audience for feedback and discussion. The visualizations to be presented may or may not have been created by the learners themselves.

Myller et al. (2009) extend Naps et al.'s engagement taxonomy by adding the following categories (p. 7, 8, the added categories are presented in bold in the following list):

- No viewing.
- Viewing.
- **Controlled viewing:** The visualization is viewed and the students control the visualization, for example, by selecting objects to inspect or by changing the speed of the animation.
- **Entering input:** The student enters input to a program or parameters to a method before or during their execution.
- Responding.
- Changing.
- **Modifying:** Modification of the visualization is carried out before it is viewed, for example, by changing source code or an input set.
- Constructing.
- Presenting.
- **Reviewing:** Visualizations are viewed for the purpose of providing comments, suggestions, and feedback on the visualization itself or on the program or algorithm.

We note, though we do not concentrate in this guide on system-level computer architecture, that simulations for system-level computer architecture that uses visualization and animation are also used in computer science education (Yehezkel et al. 2007; Taghavi et al. (2009). Yehezkel (2002), for example, describes taxonomy for visualization of computer architecture and introduce the EasyCPU environment[4] in the context of this taxonomy.

The purpose of Activities 64–70 is to let the students explore different ways by which it is possible to use animation and visualization in their future computer science classes. Activity 64 focuses on algorithm visualization; Activity 65 examines music as a kind of visualization; Activity 66 focuses on software visualization and animation; Activity 67 explores visualization-based IDEs; Activity 68 suggests how to teach sorting algorithms by YouTube dance clips; Activity 69 explores the Media Computation teaching approach; and Activity 70 summarizes what the students have worked on in Activities 64–69.

Since most of Activities 64–70 are based on the exploration of environments that use animation or visualization in some way, more than one lesson of the MTCS

---

[4] See http://csta.villanova.edu/CITIDEL/handle/10117/216.

course should be dedicated to this topic. If time constrains do not allow it, some of the exploration work should be done by the students at home as a preparation toward the lesson.

Activity 64: Algorithm Visualization

- Stage A: Increasing students' attention to visualization
  The instructor presents to the students a text that describes an algorithm with which (it is reasonable to assume) they are familiar, for example, one of the sort, search or traverse algorithms. Then, the instructor presents the following tasks:
  - Represent the given algorithm differently. (One option that will be probably suggested by the students is to represent it visually).
  - Suggest different visualization ways to present the algorithm.
    After the students' suggestions are discussed, they are asked to work on the following task:
  - Find an animation of this algorithm on the Internet and explore differences between the given text description of the algorithm and its animation. Explore these differences also from a pedagogical perspective.
- Stage B: Design a visualization-based worksheet, group work
  The students are asked to design a worksheet that teaches the said algorithm to novice computer science learners and uses the algorithm animation they found on the Web. The students can choose to develop a worksheet which is either based on the lab-first approach (see Sect. 8.8.3 above) or based on the regular approach applied in the lab after the learners were introduced to the algorithm in the class. The students should be asked also to explain each of their considerations in the design process of the worksheet.
- Stage C: Introducing Nap's (extended) Engagement Taxonomy, class discussion
  The instructor of the MTCS course:
  - Presents Nap's Engagement Taxonomy (see above) to the students (alternatively, the extended taxonomy can be presented).
  - Completes Table 8.3 together with the students with respect to general algorithm animation.
- Stage D: Examination of questions' potential contribution to learners' learning, group work
  Each group categorizes the questions it included in the worksheet designed in Stage B according to Nap's Engagement Taxonomy and examines each question's potential contribution to learners' learning of the said computer science topic. If needed, the students are asked to update or modify the questions based on this examination.

Table 8.3 Questions for each level of Nap's engagement taxonomy

Stage	Kinds of questions that are suitable to be asked in each level of engagement	How do the questions improve the understanding of the learned topic?
No viewing		
Viewing		
Responding		
Changing		
Constructing		
Presenting		

- Stage E: Summary
  In the summary of this activity, it is important to highlight that visualization and animation have the potential to promote computer science learning from several reasons. First, visualization and animation support very naturally active learning; second, they enable learners to concretize abstract concepts; and finally, learners may conceive visualization and animation as a kind of a game (see Chap. 7), and therefore their use may increase the learners' interest in computer science.
- Stage F: Homework
  - *Option 1*: The students work at home on the following activities, which offer them an additional opportunity to rethink the different usages of visualization and animation in the context of computer science education.
    - Develop the second kind of worksheet than the one you developed in the class—the worksheet is given to learners either before or after they learn the said computer science topic (see Stage B).
    - Formulate guidelines for the development of a worksheet that is based on learners' engagement with algorithm/program animation/visualization. Reflect: What considerations guided you in the formulation of these guidelines?
  - *Option 2*: Explore the algorithm visualization portal.[5] Suggest at least five usages of this website in high school computer science teaching.

Activity 65: Musical Debugging

Music is one aspect of visualization and can be used in computer science education in several ways. For example, this musical debugging activity (Lapidot and Hazzan 2005) increases learners' awareness to their debugging processes, the importance of debugging, and its relationship to learning processes. It also

---

[5] See http://algoviz.org/.

illustrates a different and non-conventional use of the computer lab (see also Sect. 5.5 about debugging processes).

- Stage A: Debugging a song melody, work in pairs with the computer
  The students are given a familiar song, a nonworking computer program of its melody, the program listing, and the melody. They are asked to debug the program and to document their actions and the strategies they used during the debugging process.

  While working on these tasks, the students debug the program, trying to make it a working program, that is, a program that plays the correct melody. Usually, they start with syntax errors, which are easier to detect and obtain a program devoid of syntax errors. At this stage, they realize that the program does not play properly the correct melody and start debugging the semantic (or logical) errors of the program.
- Stage B: Class discussion
  The discussion that follows the activity can be viewed on two levels.

  On the first level, content is addressed. The students present the errors they found in the program and describe their debugging strategies. This discussion offers an opportunity to address programming errors and their classification in different ways (e.g., syntax errors are the easiest to find vs logical errors are the most difficult to locate). When learners are asked to generalize their own strategies, the discussion focuses on general debugging strategies.

  The second level of the discussion focuses on topics such as the following ones: the importance of debugging in computer science and in computer science education, the teaching of debugging, and the place of debugging in the computer science curriculum. In this context, debugging can be addressed from the perspective of soft ideas (see Sect. 3.7).
- Stage C: Analyze one of the common debugging tools, homework
  The students are asked to choose one of the common debugging tools (such as NetBeans) and to comment on their choice from a pedagogical point of view. Another optional task would be to ask students prepare a lab worksheet for pupils where pupils will get familiar with one of the common debugging tools.
- Stage D: Prepare a debugging activity or read articles that address debugging, homework
  The students are asked to prepare another debugging activity or to read articles that address debugging (e.g., Spohrer and Soloway 1986; Perkins and Martin 1986) and to comment on them from a pedagogical point of view.

Activity 66: Software Visualization and Animation
There are computational environments, designed especially for pedagogical purposes that use visualization and animation to demonstrate how programs work. In order to discuss this issue meaningfully with the students, it is impor-

tant to let them experience first at least two such environments, as is laid out in what follows.

- Stage A: Exploration of object-oriented program visualization and animation, work in pairs
  The students are asked to explore applications which are based on visualization and animation. Two such applications are Jeliot[6] and BlueJ[7].
      The students are asked to explore these environments with one of the tutorials available on the Web. Specifically, the students can be asked to analyze these environments from a pedagogical perspective (that is, when and how these environments can be used for teaching purposes) and to think about advantages and disadvantages of these environments for learning purposes.
- Stage B: Class discussion
  The class discussion that follows this exploration should address topics such as the following: advantages and disadvantages of program visualization, the fitness of such environments to divers/all learners, concepts that their learning is supported by these environments, and misconception that such environments may create.
- Stage C: Read a paper, homework
  The students are asked to find, read, and submit a critical report on one of the research papers that deals with novice learning of the object-oriented paradigm with one of these environments (see Chap. 4 for additional details about assignments that deal with research on computer science education).

Activity 67: Visualization- and Animation-Based IDEs
Visualization- and animation-based IDEs use visualization/animation to support learners' first programming steps, by allowing learners to work with more concrete objects in their first learning steps of computer science. Here are several examples of such IDEs[8] (listed here alphabetically): Alice,[9] Greenfoot,[10] Karel J. Robot,[11] MicroWorlds,[12] Scratch,[13] Squeak,[14] Starlogo.[15] It is notable, though, that many of them are inspired by Papert's Mindstorms philosophy (Papert 1980).

---

[6] Source: http://cs.joensuu.fi/jeliot/.

[7] Source: http://www.bluej.org/download/download.html.

[8] See TOCE special issue on Initial Learning Environments (November 2010, vol. 10(4)) for a broader discussion of IDEs.

[9] Source: http://www.alice.org/.

[10] Source: http://www.greenfoot.org/.

[11] See: http://csis.pace.edu/~bergin/KarelJava2ed/Karel++JavaEdition.html.

[12] See http://www.microworlds.com/.

[13] See: http://scratch.mit.edu/.

[14] See: http://www.squeak.org/.

[15] See http://education.mit.edu/starlogo/.

The students are asked to explore these IDEs in two stages, as is described in what follows.

- Stage A: Solving a problem in one of the visualization-/animation-based IDEs, work in pairs
  The students are asked to solve a problem in each (or in selected ones) visualization or animation-based IDE that illustrates the IDEs' pedagogical approach. It is recommended that the instructor of the MTCS course gives the students specific task(s) to work on. The above websites suggest a variety of tasks that can fit for this purpose.
- Stage B: Analysis of the visualization-/animation-based IDEs, class discussion
  - What are the differences between the IDEs?
  - What are the advantages and disadvantages of each IDE?
  - What are the differences between these IDEs and algorithm animation?

Activity 68: Learning About Sorting Methods by YouTube Dances

- Stage A: Exploratory Work with the Computer (individually or in pairs)
  The students get the worksheet presented in Table 8.4, in which they are asked to watch several sorting methods as they are being demonstrated by folk dances. To support computer science educators, the title of each video is presented as well; they should be omitted when the worksheet is given to the learners.
- Stage B: Class discussion
  The class discussion that follows this exploration should address topics such as the following: what is the sorting algorithm demonstrated by each dance?, what are the differences between the algorithms?, which algorithm is easier (or more difficult) to explain to novice computer science learners and why?

Table 8.4 Dancing with numbers on YouTube worksheet

	Worksheet
	Dancing with numbers
	Watch the following YouTube links. Each video demonstrates "something" that is being done with numbers. Watched the videos and address the following questions:
1.	Which problem is solved by each dance?
2.	For each video, explain the method used in order to solve the problem.
	The links: 1) http://www.youtube.com/watch?v=Ns4TPTC8whw&feature=related 2) http://www.youtube.com/watch?v=lyZQPjUT5B4http://www.youtube.com/watch?v=lyZQPjUT5B4 3) http://www.youtube.com/watch?v=ROalU379l3U&feature=related 4) http://www.youtube.com/watch?v=XaqR3G_NVoo&feature=related

The dances performed in the following links demonstrate select sort (#1), bubble sort (#2), insert sort (#3), and merge sort (#4)

- Stage C: Homework
  Here are several options for homework tasks:
  - Design a lesson devoted to one of the sorting algorithms.
  - Search YouTube for other video clips that may be used by CS teachers in their teaching.
  - Compare types of CSE video clips you found on YouTube.

"*Media Computation* (nicknamed "MediaComp") is a contextualized approach to introducing computing using a ubiquitous theme of manipulating media. The critical characteristic of MediaComp is that students create expressive media by manipulating computational materials (like arrays and linked lists) at a lower level of abstraction. Students manipulate images by changing pixels, create sounds by iterating over samples, render linked lists into music, and create artifacts like collages, music, and digital video special effects. In so doing, the students learn computation."[16]

The MediaComp approach utilizes the advantages of visualization and animation in a different way than IDEs, such as the ones described in the previous section. That is, instead of creating a new visualization- and animation-based development environment, it uses visual elements of programming languages such as Python or Java, to teach computer science concepts.

Activity 69: Media Computation

- Option 1:
  The MediaComp approach can be addressed in the MTCS course by asking the students to prepare a teaching unit (see Chap. 11) about an introductory computer science topic (e.g., loops or two-dimensional arrays) that uses the media computation approach. While preparing the teaching unit, the students should be asked to keep reflecting and analyzing their pedagogical considerations as well as differences between the Computational Media approach and the traditional approach of computer science teaching.
- Option 2:
  The students can be referred to several textbooks that apply the Media Computation approach and to The Media Computation Teachers Website,[17] which lists these textbooks as well as additional resources.

---

[16] Source: http://coweb.cc.gatech.edu/mediaComp-teach.
[17] See http://coweb.cc.gatech.edu/mediaComp-teach.

Activity 70: Summary Work
As a summary of the topic of visualization and animation, the students can be asked to work on the following summarizing activity which integrates several aspects of the previous activities. Specifically, the students are asked to summarize all the usages of visualization and animation with which they have become familiar so far: software visualization, algorithm animation, visualization-based IDEs and the media computation approach. They can be offered to address the following topics. It is recommended to guide them to base their work on the research available on visualizations and animation in the context of computer science education (see Chap. 4).

- Differences and similarities between the different kinds of visualization and animation.
- The pedagogical purpose(s) for which each kind of visualization and animation fits.
- Types of tasks that can be given with each kind of visualization and animation tool.
- Connections between the lab-first approach and each of these applications of visualization and animation.
- Reflective essay of their future use of each application in their computer science teaching in the high school.

8.5   Using the Internet in the Teaching of Computer Science

The Internet is an inseparable learning environment in the teaching of any subject. In addition to the general arguments for integrating the Internet in learning and teaching processes, in computer science education it gets special attention since the Web by itself is largely based on computer science ideas, such as data compression, encoding, search algorithms, and data mining. Therefore, the students, as future computer science teachers, should be familiar with this environment as a learning environment in general, and with some computer science ideas that are implemented by the Web, in particular.

The Internet enables a constructivist-based learning environment in which learners are active. In this chapter, we have seen so far one use of the Internet for the learning of computer science in the context of algorithm animations. Activities 71–73 focus on the following ways which use the Internet for the learning of computer science:

- Information gathering
- Exploration of the Internet through the computer science lens
- Distance learning

Activity 56 on flipped classroom (Sect. 7.3) is another use of the Internet for computer science teaching. Other examples are Google Trends as well as other technologies (such as, Smartphone) can be explored in the MTCS course as well.

Activity 71: The Internet as an Information Resource
The students are asked to design a lesson, which is based on the Internet as an
information resource, on some computer science topic that is learned in the
high school, for example, the history of computer science or different kinds
of sorting algorithms.
   After the students designed this lesson and their suggestions are presented
and discussed in class, a discussion takes place that focuses on what computer
science topics fit to be learned by this teaching approach.

Activity 72: Exploring the Internet Through the Computer Science Lens

- Stage A: Analysis of Web applications, group work
  The students are asked to explore different Web applications from the perspective of computer science. In other words, the students explore what
  computer science ideas are implemented and used in different web-based
  applications with which they are familiar. They can be directed to focus
  on either the software design (e.g., the design of a profile in one of the
  online social networks) and/or on an algorithm used by an online application (e.g., search in one of the search engine).
- Stage B: Presentation of the group works
  While the students present their products in front of the class, it is recommended to check whether the computer science topics selected for exploration can be integrated into the high school curriculum that the students will
  teach as high school computer science teachers. It is reasonable to assume
  that even if not all topics can be merged as a whole into the high school
  curriculum, some of them can be integrated partially. It is important to
  remember, though, that even just mentioning in high school computer science classes the applications of computer science ideas in some real Web
  applications, with which the pupils are familiar, can increase the pupils'
  motivation to study computer science.

Distance learning is a vast topic that is still explored. Many questions about how to
apply distance learning effectively are still open. However, similar to face-to-face
teaching situations, with respect to which it is clear that active learning promotes
learners' understanding (see Chap. 2), it is clear that in online learning environments in general and distance learning environments in particular, learners should
be active. This assertion is based on the fact that in distance learning situations, the

face-to-face social interaction offered to learners in traditional teaching processes should be substituted with another mechanism that enhances learners' engagement.

We note that the activities presented in this guide are based on face-to-face learning and teaching situations and can be adjusted for distance learning situations, by keeping the active-learning-based teaching model (Chap. 2) and facilitating the different stages of the model using an online platform.

Activity 73: Distance Learning

- Stage A: Learning a new computer science topic, individual work
  The instructor of the MTCS course selects an online lesson about a computer science topic with which the students are probably not familiar. The instructor asks the students to learn the lesson and in parallel to reflect on their learning process.
- Stage B: Class discussion
  Students' reflections are shared in front of the class. It is important to highlight different aspects of the students' reflection, such as cognitive, affective, and social ones.
- Stage C: Integration of distance learning elements into the computer science curriculum, homework
  The students are asked to review the high school computer science curriculum they are going to teach in the future and to suggest specific places in which they will be able to integrate some distance learning elements. Each decision should be explained and its contribution to learners' learning of computer science should be outlined.

References

Ben-Bassat Levy R, Ben-Ari M (2007) We work so hard and they don't use it: acceptance of software tools by teachers. Proceedings of the 12th Annual SIGCSE Conference on Innovation and Technology in Computer Science Education, Dundee, Scotland, UK; 246–250

Ben-Bassat Levy R, Ben-Ari M (2008) Perceived behavior control and its influence on the adoption of software tools. Proceedings of the 13th Annual SIGCSE Conference on Innovation and Technology in Computer Science Education, Madrid, Spain: 169–173

Friedman DP, Felleisen M (1986) The little LISPer. Science Research Associates, Inc., Palo Alto

Knox D, Wolz U, Joyce D et al (1996) Use of laboratories in computer science education: guidelines for good practice. Report of the Working Group on Computing Laboratories. Integrating Tech. into C.S.E. 6/96 Barcelona, Spain

Lapidot T, Hazzan O (2005) Song debugging: merging content and pedagogy in Computer Science education. Inroads—the SIGCSE Bull. 37(4):79–83

Lee MJ (2013) How can a social debugging game effectively teach computer programming concepts? ICER'13, San Diego CA USA: 181–182

Ma J, Nickerson JV (2006) Hands-on, simulated, and remote laboratories: a comparative literature review. ACM Comput Surv 38(3), Article 7

Myller N, Bednarik R, Sutinen E et al (2009) Extending the engagement taxonomy: software visualization and collaborative learning. Trans Comput Educ 9(1):1–27

Naps T, Rößling G, Almstrum V et al (2003) Exploring the role of visualization and engagement in computer science education. ACM SIGCSE Bull 35(2):131–152

Nersessian NJ (1991) Conceptual change in science and in science education. In Matthews MR (ed) History, philosophy, and science teaching. OISE Press, Toronto, pp 133–148

Papert S (1980) Mindstorms: children, computers and powerful ideas. Basic Books Inc., New York

Paz T (2006) Introduction to computer science in Java, worksheet collection (Hebrew). Oranim Academic College of Education, Israel, pp 33–36

Perkins DN, Martin F (1986) Fragile knowledge and neglected strategies in novice programmers. In Soloway E, Iyengar S. (eds) Empirical studies of programmers. Ablex Publishing Co., Norwood, pp 213–229

Petre M (2011) Online experimentation. ACM Inroads 2(1):18–19.

Price B, Baecker R, Small I (1998) An introduction to software visualization. In Stasko J, Domingue J, Brown, M, Price B (eds) Software visualization. MIT Press, Cambridge, pp 3–27

Rössling GJ, Velázquez-Iturbide JÁ (2009) Editorial: program and algorithm visualization in education. ACM Trans on Comput Edu (TOCE) 9(2):6–11

Shaffer CA, Cooper ML, Alon AJD et al (2010) Algorithm visualization: the state of the field. ACM Trans Comput Educ 9:1–22

Spohrer JG, Soloway E (1986) Analyzing the high frequency bugs in novice programs. In Soloway E, Iyengar S (eds) Empirical studies of programmers. Ablex, Norwood, pp 230–251

Taghavi T, Thompson M, Pimentel AD (2009) Visualization of computer architecture simulation data for system-level design space exploration. Embedded computer systems: architectures, modeling, and simulation. Lect Notes Comput Sci 5657(2009):149–160

Urquiza-Fuentes J, Velázquez-Iturbide JÁ (2009) A survey of successful evaluations of program visualization and algorithm animation systems. ACM Trans Comput Edu (TOCE) 9(2):1–21

Walker HM (2011). A lab-based approach for introductory computing that emphasizes collaboration. Proceedings. of CSERC '11 Computer Science Education Research Conference, Heerlen, Netherlands, 21–31

Yehezkel C (2002) A taxonomy of computer architecture visualization. ACM SIGCSE Bull 34(3):101–105

Yehezkel C, Ben-Ari M, Dreyfus T (2007) The contribution of visualization to learning computer architecture. Comput Sci Educ 17(2):117–127

Types of Questions in Computer Science Education

9

Abstract

It explores and discusses different types of questions that computer science (CS) educators (middle and high school teachers as well as university instructors) can use in different teaching situations and processes: in the classroom, in the computer lab, as homework, or in exams. The chapter discusses also keywords that appear in problem-solving questions which reflect the need to apply high-order cognitive skills by learners when answering these questions. The chapter lays out the advantages of using a variety of question types both for learners and teachers, and focuses on the design process of different question types. Though the types of questions presented are mainly related to programming assignments, most of them are suitable also for other CS contents.

9.1 Introduction

The main target of this chapter is to enrich computer science (CS) educators' toolbox with respect to the design of different types of questions and to illuminate the important role of CS educators in exposing their pupils/students to different types of questions throughout the learning-teaching processes. Learners' work and exploration of different types of questions and their variations deepen their understanding of the learned CS concepts, refine their understanding of complex concepts, and let them acquire different cognitive skills. Such an experience also provides learners with the opportunity to express their knowledge in different forms. Further, the use of a variety of question types sometimes provides intellectual challenges and maintains learners' concentration, interest, and motivation.

© Springer-Verlag London Limited 2014
O. Hazzan et al., *Guide to Teaching Computer Science*,
DOI 10.1007/978-1-4471-6630-6_9

This teacher's role should be delivered to the prospective CS teachers, and therefore, it is recommended to use/mention/practice a variety of different types of questions throughout the Methods of Teaching Computer Science (MTCS) course in different opportunities and contexts. Clearly, instructors of the MTCS course should also apply the same pedagogical principles and vary the types of tasks and questions they use while discussing the teaching of CS with the prospective CS teachers.

One of the common problem-solving scenarios in CS education starts with the presentation of an open problem that describes some story, continues with the problem analysis and planning of its solution, and ends with the presentation of the solution as an algorithm either in pseudocode or a specific programming language. It is important, however, that CS teachers be aware of the fact that additional types of questions exist.

Several pedagogical targets can be achieved by the integration of different types of questions in the teaching of CS, as is laid out in what follows:

1. Different types of questions enable to illuminate different aspects of the learned content.
2. The integration of different types of questions throughout the teaching process helps maintain the students' interest, attention, and curiosity.
3. Different types of questions enable teachers to vary their teaching tools.
4. Different types of questions require the students to use different cognitive skills—mental abilities we use while thinking, learning, and studying. This target is important to enable each learner express his or her unique individual cognitive skills, to articulate his or her knowledge in his or her unique way, and to develop and enrich one's cognitive skills.

Cognition is the process of thought and it can be analyzed from different perspectives. For example, in psychology or philosophy, the concept of cognition is closely related to abstract concepts, such as mind, reasoning, perception, intelligence, and learning, all of which describe the mind capabilities. The field of cognition studies specifies mental processes, such as comprehension, inference, decision making, planning, and learning. With respect to CS, we are all familiar with the advanced cognitive skills of abstraction, generalization, concretization, and meta-reasoning.

Research works in CS education deal with different types of questions from the cognitive perspective. For example, Thompson et al. (2008) reviewed the work that has been done throughout the last 10 years with respect to the application of Bloom's taxonomy (Bloom et al. 1956) to CS course design, evaluation, and assessments, and provided an interpretation of the taxonomy that can be applied to introductory programming exams. Their interpretation focuses on the cognitive skill involved in addressing several types of questions. Jones et al. (2009) focus on written examinations. They determine the difficulty level of each question by keyword/s found in the question, and present cross-analysis that addresses student performance, cognitive skills, and learning outcomes. Other kind of works that combine CS questions and cognition relates to automata systems of question-answering processes (see, e.g., Pomerantz 2002; Yang et al. 2008).

We first concentrate on question *patterns* and present 12 types of questions that CS educators can use (Sect. 9.2). For each type of question, we lay out several variations. Clearly, additional types of questions, as well as the combination of different types of questions, exist and can be developed and used by CS teachers. Section 9.3 presents keywords of problem-solving questions. Then, Sect. 9.4 presents three general kinds of questions (story questions, closed questions, and unsolvable questions). Section 9.5 describes how the different types of questions can be assimilated to different CS contents and demonstrates these assimilations in the context of Automata Theory. In Sect. 9.6, we present guidelines on how to develop questions to be used in a CS class. We suggest several course activities to be facilitated in the MTCS course in order to expose the prospective CS teachers to this topic.

9.2   Types of Questions

This section presents 12 types of questions and suggests how they can be used in CS teaching processes. Emphasis is placed on the pedagogical approaches they represent and on cognition considerations.

Each type of question is presented according to the following pattern: *classification* of the type of question reflected by its title; short *description* of the specific type of question; a concrete *example* or an *example of a general pattern* that demonstrates the said type of question; different *variations* of the discussed type of question; and a short pedagogical and cognitive *discussion* about the said type of question. Since our purpose here is to present the variety of questions to future high-school prospective teachers, most of the examples are quite simple. Clearly, for each type of question, it is possible to develop a range of questions on different complexity levels both from the algorithmic and the cognitive points of view. In addition, with respect to each type of question, it is possible to present additional variations that require different cognitive skills.

We add several remarks about the actual use of this collection of questions in the actual teaching of CS situations:

- The order of the presentation of the 12 types of questions in this chapter is arbitrary; no specific rule is applied for their ordering.
- No specific rules or guidelines can be formulated with respect to the order by which it is recommended to preset the different types of questions in the class; each CS educator should select the appropriate type of question and its complexity level according to the specific learners' characteristics and the specific teaching situations.
- The suggested types of questions are sometimes overlapping each other; this point is addressed when it is relevant.
- A question can contain several types of questions in its different subtasks; this point is illustrated by a particular example in Sect. 9.2.13 "Combining several types of questions."

- In general, questions can be divided into two types: Pure algorithmic tasks and story-based algorithmic tasks; this perspective is discussed explicitly in Sect. 9.4.1 "Story questions."
- The types of questions presented in this section are mainly programming assignments; most of them, however, can be easily assimilated for other CS contents.

9.2.1   Type1. Development of a Solution

Description: A development question presents an open problem in which learners are required to develop their solution to a given problem. The solution can be expressed by a descriptive algorithm, pseudo-code or a program in a specific programming language.

Example: Write a method that returns the number of (integer) divisors for a given integer n.

Variations: A development question can ask for, for example: (a) a single method (as the presented example); (b) a sequence of tasks to be performed; (c) a complete program; or (d) a method with a specific efficiency (in the example presented it can be $O(\sqrt{n})$).

Discussion: This type of questions can be solved in different ways. In some cases, the differences are not meaningful; in other cases, the different solutions represent different algorithmic approaches.

Variation (d) is not a fully open question since learners are asked to address a specific constraint—specific efficiency, and cannot develop any solution for the problem; therefore, this variation requires wider range of considerations than the other ones and is considered a harder question than the other variations.

9.2.2   Type2. Development a Solution That Uses a Given Module

Description: In this case, the development question relates to a pre-prepared module and asks learners to present a solution to a given problem while considering and using a given module. A documentation of the module is included in the question and the student must use it in their solution.

Example: Write a method that returns an integer number between $1-n$ that has the largest number of divisors for a given integer n. Use the method *numberOfDividers*(n), that returns the number of divisors for a given integer n.

Variations: A development question that relates to a given module can be presented, among other ways, in one of the following forms: (a) write an instruction that invokes a given method; (b) write a method that uses a given method (as in the given example); (c) write a method that uses a given module a specified number of times; (d) write a method that uses several different given methods; (e) questions in which the given module is not a method, but rather it is, for example, a specified data structure or a specified class.

Discussion: The fact that students should relate to a given module influences the development process of the solution. For example, in the case of a given method, as in the above example, the learner has to suit the developed method to a specific sub-task that the given method implements. This type of questions is considered harder than Type1 questions because in this case learners need to meet a constraint—the use of the given subtask.

9.2.3   Type3. Tracing a Given Solution

Description: A given code is presented and the learners are asked to track the code execution.

Example of a general pattern: Present a tracing table that follows the execution of a given method. The table should include a column for each variable and for the code output.

Variations: A tracing question can ask to follow, for example: (a) a complete program; (b) a single method; (c) a recursive method; (d) object creation. In addition, the following instructions can be used in each of the above variations: (1) follow the code execution according to a given input; (2) follow the code execution when learners choose the input; (3) follow the code execution according to several different specified inputs which are selected in a way that guides the learners to find what the given code performs; (4) find different sets of inputs so that each set represents a different flow by which the code is executed; (5) find a set of inputs that yields a specific output.

Discussion: Variations (1)–(3) can be considered as closed questions. The learner is required to trace a given code with a specified (given or chosen) input, and there is only one correct solution. Variations (4) or (5) require learners to apply deeper considerations and to examine the presented code from a higher level of abstraction. In these cases, it is not sufficient to understand different instructions; rather, they require code analysis—what the purpose of the code is and how it is achieved. Clearly, more advanced cognitive skills are needed in order to address meaningfully these variations.

9.2.4   Type4. Analysis of Code Execution

Description: A given code is presented and learners are asked to analyze specific aspects of the code execution.

Example of a general pattern: Look at the given code that includes a loop and answer the following questions:

1. For what values of x and y the loop is not executed at all?
2. For what values of x and y the loop is executed exactly one time?
3. For what values of x and y the loop will never terminate?

Variations: Variations (4) and (5) of Type3 questions—tracing a given solution—can be viewed also as variations of this type of questions.

Discussion: In this type of question, the learner is required to analyze the code execution and to understand it as a whole. Specifically, in order to solve such questions, learners should exhibit mainly two cognitive skills: understanding programming structures and understanding the logic of a given code. Therefore, a higher level of thinking is needed to solve such questions than that needed for solving a tracing question; accordingly, this type of question is considered harder than the "tracing a given code" type of question.

9.2.5 Type5. Finding the Purpose of a Given Solution

Description: A given solution to an unknown problem is presented and the learners are asked to state the purpose of the solution, that is, to determine what problem it solves.

Example of a general pattern: Look at the given method and write the method target, that is, what is the problem that the method solves?

Variations: A "finding the purpose of a given solution" question can relate to either: (a) a sequence of instructions; (b) a single method (like in the presented example of a general pattern); (c) a full program; (d) a class.

Discussion: This type of questions is considered harder than tasks that ask to develop a solution for a given problem. For solving this type of questions, a set of cognitive skills is required. Specifically, in addition to the understanding of the code execution and the ability to trace it, one should comprehend someone else's way of thinking.

To help students solve this type of questions, a question can contain scaffolding subquestions. For example, a question can include several tracing subquestions (Type3 questions), which aim to guide the students to discover the purpose of the code.

9.2.6 Type6. Examination of the Correctness of a Given Solution

Description: A given problem and its solution are presented. The student is asked to determine whether the given solution solves the given problem correctly.

Example: The following method was written by a student as a solution for the following problem:

Write a method that returns *true* if all values of a given array of integers are equals; otherwise, it returns *false*.

Is the method correct?

```
public static boolean equalValues(int[] arr) {
            for (int i=0; i<arr.length; i=i+2) {
                if (arr[i] != arr[i+1])
                    return false;
            }
        return true;
}
```

Variations: This type of question can be presented in different forms: (a) determine whether a given solution to a given problem is correct (as in the example presented); (b) check if a given solution to a given problem is correct and explain your answer; (c) if the given solution is incorrect, give an example of an input that shows it; (d) if the given solution is incorrect, give an example of an input that presents a correct output and, therefore, may mislead one to conclude that the given solution is correct; (e) if the given solution is incorrect, correct the solution by introducing the minimal required changes (without this restriction, students may present a totally different solution); (f) the presented solution can contain more than one mistake, and the question can state it explicitly or not.

Additional variations of this type of questions may address syntactic mistakes. It is suitable to present such variations while introducing new instructions or data structures. It is not recommended, however, to use these variations in more advanced stages since they do not indicate learners' understanding of the *algorithmic* problem, and, further, they do not contribute meaningfully to learners' understanding since, in fact, the compiler directs how to debug such mistakes.

Discussion: In order to solve this type of questions, students should apply algorithmic thinking and logical skills. Here, as in Type5 questions, students should analyze a solution that may not fit their own way of thinking had they been asked to develop a solution. However, since the purpose of the solution is given, these tasks are considered easier than Type5 questions.

In the example presented, the two minimal required corrections are: (1) change the increment of variable *i* to *1* (instead of *2*); (2) change the range of variable *i* to be *i<arr. length-1*. Correction (1) is based on a logical consideration, while correction (2) addresses the array index, which is a more technical consideration.

9.2.7   Type7. Completion of a Given Solution

Description: A given problem and an incomplete solution of the given problem, in which some of the instructions are missing, are presented to the learners. The learners are asked to complete the missing instructions, so that the solution will solve the problem correctly.

Example: The following method was written by a student as a solution for the following problem: Write a method that for a given array of integers returns the number of array elements that are bigger than their two neighbors (the previous element and the subsequent element in the array).

```
public static int numberOfBigger(int[] arr) {

    _____ ;
    for (int i=_____ ; i<_____ ; i++) {
        if ( _____ )

            _____ ;
    }
    return _____ ;
}
```

Variations: "A completion a given solution" question can be varied by changing the number of the missing instructions. This number should be determined by taking into the consideration that it may affect the difficulty and complexity of the question.

In general, the missing instructions can relate to one or more aspects of the algorithm and the teacher should consider whether to focus on one or more aspects. For example, if the teacher's target is to focus on the use of a *boolean* flag, the missing instructions should be only those that relate to this flag; if the target is to focus on the loop limits, the limits should be the missing parts and, sometimes, also the increment of the loop control variable.

In the example presented, the missing instructions are related to three aspects: the counter control (initialization, increment, and return); the range of the loop (the first and the last array elements should not be accessed in the loop because they do not have two neighbors); and the specific condition to be checked.

Discussion: This type of question, as well, requires students to understand the logic of the given solution, when the actual question difficulty is determined according to students' level and stage of learning. Still, with respect to each CS subject, relatively simple "completion of a given solution" questions exist, where the understanding of the logic of the solution is straightforward. At the same time, however, there are more challenging questions and the teacher should be aware of this complexity. For example, asking to complete instructions in a given bubble sort algorithm, in which meaningful instructions are missing, without introducing the rational of this sorting approach, is considered a difficult question.

9.2.8   Type8. Instruction Manipulations

Description: A problem and its solution are given. Students are asked to address different manipulations performed on the solution.

Example: The following method executes a variation of the selection sort.

```
public static void selectionSort(int[] arr) {
    int p, temp;
(1)    for (int i=0; i<arr.length-1; i++) {
(2)        p = i;
(3)        for (int j=i+1; j<arr.length; j++) {
            if (arr[j]<arr[p])
                p = j;
        }
(4)        if (p != i) {
            temp = arr[i];
            arr[i] = arr[p];
            arr[p] = temp;
        }
    }
}
```

Answer the following questions and explain your answers:

1. Will the algorithm correctness be effected if the instruction marked by (2) is removed and the instruction marked by (3) is replaced with the following instruction:
   (3) *for (int j = i; j < arr.length; j + +) {*
2. Will the algorithm correctness be affected if the instruction marked by (4) is removed and the contents of the two array elements are swapped anyway?
3. Will the algorithm correctness be affected if *all* the body of the loop marked by (3) is replaced with the following instructions?

   (3)   *for (int j=i; j<arr.length; j++) {*

Variations: An "instruction manipulations" question can be implemented by the (a) addition of instructions; (b) removal of instructions; (c) changing instructions; (d) replacement of instructions. The question itself can address the target of a specific code or the tracing of the changed code and the examination of differences between outputs.

Discussion: Questions that manipulate a given solution enable to concentrate on meaningful aspects of the algorithm. In the teaching process, a discussion on such manipulations can clarify the essence of a given solution as well as other CS topics. We illustrate this idea by the concept of generalization. Students can be instructed to change a given method slightly, in a way that generalizes the original method and solves a broader task. For example, a method that sorts an array can be slightly changed in order to sort a section of the array between two given indices. After the change, the method can sort different sections of an array, as well as the entire array (with the indices 0 and the array length—1).

9.2.9   Type9. Efficiency Estimation

Description: Students are asked to estimate the efficiency of a given solution.

Example of a general pattern: Estimate the efficiency of the presented method in terms of *big O*. Explain your estimation.

Variations: This type of questions can represent different levels of cognitive complexity, as is described in what follows. (a) A general pattern that enables to discuss efficiency in early stages of CS learning: Focus on the loop in the given method: How many times is the loop executed?; (b) Estimate the efficiency of a specific method (as in the example of a general pattern); (c) Estimate the efficiency of a method that invokes another method, when the efficiency of the invoked method is taken into the consideration; (d) Compare the efficiency of different methods that solves the same task; (e) Estimate the efficiency of a recursive method; (f) Develop of a solution to a given problem with a specific efficiency.

Discussion: Instructors should not wait till they teach complicated algorithms in order to teach the concept of efficiency; alternatively, questions that deal with efficiency can be integrated from early stages of teaching and learning CS. For example, questions, such as the one presented in variation (a), hints at the seeds of the concept of efficiency that, in general, is considered to be abstract and difficult for understanding.

Note that variation (f) is actually a "development of a solution" (Type1 question) with a restriction about its efficiency, which requires learners not to be satisfied when finding an algorithmic idea that solves the given problem; rather, they should estimate its efficiency and if it does not fit the restriction mentioned in the question, they should look for another solution or improve the solution they found.

9.2.10   Type10. Question Design

Description: Students are asked to design a question.

Example1: Design a question that checks learners' understanding of the sort-merge algorithm.

Learners' answer to this question can be based, for example, on tracing regular and/or extreme cases of the input to this algorithm.

Example2: Design a question in such a way that its solution uses a method that finds the most frequent value in an array. An example for such a question is:

Pupils' grades in a CS test are given. Write a method that prints the most frequent grade in this class.

Variations: A design question can relate, among other options, also to the design of (a) additional tasks in a given question that clarify extreme cases; (b) a question that its solution should use a given method (as Example2 is); (c) a question that intends to check the understanding of a specific concept or algorithmic idea (as Example1 is); (d) a whole test or worksheet that examines a specific CS concept/topic.

Discussion: This type of questions invites learners to adopt a different point of view. In addition to the experience learners gain by examining a question from the educator's point of view, this kind of questions encourages them to scrutinize the learned concepts, to reflect on what they learned, and, by doing so, also to evaluate their own understanding. In addition, the design of the questions is a kind of active learning that encourages creativity.

9.2.11 Type11. Programming Style Questions

Description: Learners are asked to examine the programming style of different solutions presented for the same task.

Example of a general pattern: Look at the given collection of *correct* solutions for a given problem. Examine the solutions and state which of them, in your opinion, is the best solution. Explain your choice.

Variations: The different solutions for the given problem can differ in one or more aspect(s), according to the teacher decision, such as (a) different kind of loops; (b) the need to use an array for the solution; (c) different algorithmic approaches (e.g., given two correct solutions to a given problem, the learners should decide which one is a better solution and explain why). If the teacher decides to integrate in the question several aspects, the different aspects can be presented explicitly in the question when learners are asked to analyze the solutions according to each aspect.

Discussion: This type of question enables to foster a discussion about different aspects of programming style, which in turn increases learners' awareness to these aspects.

9.2.12 Type12. Transformation of a Solution

Description: A problem and its solution are presented to the learners in a specific programming approach, programming language, or programming paradigm. The learners' task is to transform the solution into a different programming approach, a different programming language, or a different programming paradigm.

Example of a general pattern1: The presented loop is implemented by a *while* loop. Implement it by a *for* loop.

Example of a general pattern2: The method presented is implemented by a *while* loop. Implement it by a recursive method that achieves the same target.

Example of a general pattern3: The following method sorts an array of integers in the *imperative* programming paradigm. Implement it in the functional programming paradigm.

Variations: The different transformations presented in this kind of questions can be (a) between programming paradigms (as in Example of a general pattern3). This variation can be carried out only after the two said programming paradigms were learned; (b) within the same programming paradigm but between programming languages; (c) within the same programming language but between structures (as in the example of general pattern1 or, for example, from a nested *if* statement to *switch-case* statement); (d) within the same programming language but between different algorithmic approaches (as in the example of general pattern2); (e) between differ-ent representations, for example, from pseudo-code to any formal language. This variation, however, does not foster problem-solving skills and does not involve meaningful CS concepts. It can serve, however, for practicing different kinds of algorithm representations.

Discussion: The focus in this type of questions should be placed on concep-tual aspects, rather than on syntactic aspects. By conceptual aspects we refer, for example, to problem analysis according to two different programming paradigms or the transformation of a sequential solution into a recursive solution in the same programming language.

In a similar way to the "programming style questions" (Type11), "transforma-tion of a solution" questions enable to concentrate on core CS concepts. In addition, such questions lead students to explore different problem-solving approaches. It should be remembered, though, that since this type of questions demands skills of high level of abstraction, it does not necessary fit all learners.

9.2.13 Combining Several Types of Questions

Though the above types of questions attempt to classify CS questions, in most cas-es, questions either combine several types of questions (as the following example illustrates) or cannot be classified at all (see Activity 75).

Example: The target of the following two methods is to determine whether an integer number *n* is a prime number or not.

```
                    public static boolean prime(int n) {
    Method A              for (int i=2; i<n; i++) {
                             if (n % i == 0)
                                 return false;
                         }
                         return true;
                    }
```

```
                    public static boolean prime(int n) {
                         if (n % 2 == 0)
                             return false;
    Method B              for (int i=3; i<n; i=i+2) {
                             if (n % i == 0)
                                 return false;
                         }
                         return true;
                    }
```

Here is a list of questions of different types that can be asked with respect to these methods separately or in any combination according to the teacher's pedagogical purposes.

1. *Type3. Tracing a given solution*: Trace each method when *n* is *19*.
2. *Type4. Analysis of code execution*:

-   For each method, determine how many times the loop is executed for *n = 19*.
-   Find a value of *n*, for which the loop in Method B is executed ten times. Is there only one answer?

3. *Type5. Finding the purpose of a given solution*: What is the purpose of each method? (can be asked if the problem is not indicated).
4. *Type6. Examination of the correctness of a given solution*: Check the correctness of the two solutions. Do they solve the problem correctly?
5. *Type9. Efficiency estimation*: What is the efficiency of each method?
6. *Type6. Correctness; Type8. Instruction manipulations; Type9. Efficiency*:

-   If you change the upper loop limit in Method B to be *n/2* (instead of *n*), is the solution still correct? If it is, what is the method efficiency after the change?
-   If you change the loop limit in Method B to be $\sqrt{n}$ (instead of *n*), is the solution still correct? If it is, what is the method efficiency after the change?

Activities 74 and 75 focus on the above collection of question types. The activities can be facilitated in different opportunities throughout the MTCS course, not necessarily only when the topic types of questions is at the focus of the discussion. The important thing is to keep mentioning that a variety of questions exists in CS education and keep emphasizing the importance of its use. For example, when students are asked to prepare questions and activities for CS learners on a specific CS topic, this variety of questions should be reminded and the students should be asked to use it.

Activity 74: Question Classification
This activity is based on three stages: individual work, group work, and a class discussion.

A collection of questions is given to the students together with a list of types of questions. The students are asked to classify the questions according to their types. Another option is to ask the students to classify the questions according to their criteria (and not to distribute the list of types of questions).

They can be asked first to work on the classification individually, and then to discuss their classification in small groups, trying to reach an agreement on one classification. Obviously, a question can be classified into two or more types of questions, but such cases, in fact, just sharpen the students' thinking about the nature of CS questions. After the group work, a discussion in the course plenum takes place in which the classifications are presented and their rationales are articulated.

We note that in most CS textbooks at the end of each chapter, a set of questions is usually presented and the questions to be sorted can be taken from there. In this case, it is relevant to increase the students' awareness to the fact that they should examine also these sets of questions while choosing a text book to be used in their future filed work.

Activity 75: Classification of Nonsimple Questions
The students get a worksheet with patterns of questions, and are asked to work in groups and for each pattern, either to indicate its type (by using the list of types) or state that no type of questions fits it. In the second case, students are asked to formulate a new type of question and to indicate its pedagogical targets.

Table 9.1 presents a worksheet about the classification of question patterns.

Table 9.1 Worksheet about the classification of nonsimple questions

Worksheet—question classification
In what follows, a list of questions patterns is presented
For each question pattern determine its type; if you cannot find a suitable type, formulate a new type that captures its essence and indicate the pedagogical purposes of this type of questions
1. A trace table is given. Write a sequence of instructions that produces this trace table
2. In the computer lab: A given program contains a new built-in function called doSomething. Investigate the purpose of this function and formulate it. Document your investigation process
3. In what follows, several runtime errors are presented. For each of them, write a program that during its running, the error is received
4. A narrative-algorithmic question is presented. Choose the most appropriate data structure needed to solve the question, and explain your choice
5. In the computer lab: Play a given game and speculate what classes/data structures/functions used for its programming

9.3   Problem-Solving Questions[1]

The centrality of the problem-solving process was emphasized in Chap. 5, in which we offered different pedagogical tools to help learners acquire problem-solving strategies and gain experience in these skills.

---

[1] Based on Ragonis and Shilo (2013).

Table 9.2 Problem-solving question categories and keywords

Category	Keywords	Interpretation
1. Address/define criteria	Address/apply/note/mention/ specify/indicate/sort/mark	Address and apply different kinds of criteria and attribute the criteria to a list of elements; define criteria
2. Argue and justify	Argue/state/assert/ determine, follow by justification: explain/argue/prove/justify/ demonstrate/illustrate/clarify	State opinion and further establish the claim using any kind of justification
3. Analyze	analyze/examine/investigate/ explore	Identify and analyze the meaning or significance of components and factors
4. Compare	Compare/classify	Compare different objects/issues by applying principles, and observing from different view points; generalize insights
5. Complete	Complete/add	Complete or add components to a given structure according to detailed requirements
6. Convert	Convert/represent in different forms/modify/adjust/ change/transform	Convert a given paragraph/section/ clause according to specified, meaningful qualitative -related instructions (not technical translation)
7. Discover	Discover/identify/find out/ say what	Discover a phenomenon/indicate an occurrence/find out the purpose/ identify components and the relations between them
8. Develop	Develop/compose/write/create new elements	Develop a new component/write a new module
9. Integrate	Integrate/order/arrange/ merge/combine	Integrate some given components to a new structure

Boyer et al. (2010) present recommended questions for instructors to use in class discussions during question-answer sessions, in order to promote students' problem-solving skills in CS classes. The main issue emphasized is the debate climate, in which answers are not given and questions are posed to encourage meaningful thinking. In a thorough analysis of problem-solving questions, based on an extensive review of textbooks, worksheets and exams, Ragonis and Shilo (2013) investigated keywords in questions. They especially looked for questions that their target, as characterized by the authors, is to examine high-order thinking skills.

The keywords used in problem-solving questions were grouped into nine categories, by the cognitive resources learner needs to apply in order to solve a question. Each category was then represented by several keywords. Table 9.2 displays the categories, the keywords that express each category and a short interpretation of their meaning and usage. The categorization reflected different point of view than the known cognitive taxonomies (e.g., Bloom taxonomy 1956; Biggs and Collis SOLO taxonomy 1982) .

Examples of questions for each of the nine categories are presented in the Ragonis and Shilo (2013).

Activity 76: Test Analysis

This activity can be facilitated in the context of any topic discussed in the course. As preparation to the activity, the course instructor collects examples of tests to be used in the group work. Also, Table 9.2 is distributed to the students.

The students are divided into groups and are asked to analyze the test questions by examining the cognitive skills a learner needs to apply in order to solve the questions. In this activity, students are asked to apply their critical thinking: They can add keywords to the categories and to discuss the differences between "problem-solving questions" to "non problem-solving question." They can also add a new category.

Discussion can be then facilitated in the course plenary on whether a specific content influences cognitive demands or not.

9.4 Kinds of Questions

In this section, we present three kinds of questions, which represent a more global question classification. The first two kinds—*story questions* and *closed questions*—can be applied with respect to most of the 12 types of questions presented in Sect. 9.2. The third kind of question addresses *unsolved problems*.

9.4.1 Story Questions

Story questions presented to CS learners can be divided literally into two main kinds: pure-algorithmic tasks and narrative-algorithmic tasks. *Pure-algorithmic* tasks are problems that directly and explicitly address the program structures and variables, and present the task by using this terminology. *Narrative-algorithmic* tasks are problems that neither relate to program structures nor to its variables; rather, in this kind of questions, the problem to be solved is embedded in a story and in order to solve the problem, learners should recognize both what is given and what the target of the problem is. Specifically, learners should decide which elements included in the question formulation are relevant and which ones are irrelevant, and based on this decision, to solve the task. Most of the examples presented in the list of types of questions (Sect. 9.2) are pure-algorithmic tasks where the problem target is presented explicitly.

Table 9.3 presents three story tasks in two ways: as pure-algorithmic tasks and as narrative-algorithmic tasks.

It is important to address these two kinds of story questions in the MTCS course, not only because they are different, but also because they require different problem-solving skills. A pure-algorithmic question indicates specifically the task to

Table 9.3 Tasks as pure-algorithmic tasks and as narrative-algorithmic tasks

The task	Pure-algorithmic formulation	Narrative-algorithmic formulation
Find the maximum of a list of numbers	Write a method that returns the maximum value of a given list of integers	In a sport competition, 5 classes of 30 pupils each participates in two jumping competitions. Write a program that displays for each class the best result in each of the two jumping competitions for given two results of each student
Checks whether a given array is sorted	Write a method that returns true if a given array is sorted; otherwise, it returns false	A teacher wishes to encourage his or her pupils, and to give them a written recognition if their grades are improved in each test. Write a method to determine whether a given student deserves the recognition based on his list of grades
Change characters to their successive characters according to the Unicode table	Write a method that changes a given array of characters in a way that replaces each character with its successive character according to the Unicode table	A message that should be sent between financial partners should be encoded. The message includes words, spaces, and dots. Write a method that returns a coded message in which each letter of the given message (String) is replaced by its successive letter in the alphabetical order. The letter "Z" will be replaced with the letter "A." Spaces and dots should not be changed

be solved; in narrative-algorithmic tasks, students should discover the task to be solved. Since in the real world, most problems are based on narratives, the ability to solve of narrative-algorithmic tasks is an important skill that CS learners should acquire. It should be remembered, though, that these questions are more complicated.

Accordingly, when teaching a new CS content, story questions should be addressed in several stages: (1) present a general story that embeds the new learned topic, so that the class gains the essence and target of the new topic; (2) focus for a while on pure-algorithmic questions, to allow a gradual knowledge construction process of the new tool or structure; (3) integrate narrative questions in the continuation of the teaching process.

9.4.2 Closed Questions

The common interpretation of a *closed question* is a question that is presented together with a list of possible answers and the learners' task is to choose the correct answer from this list. The frequent types of closed questions are multiple-choice questions or true/false questions. It should be remembered, though, that in fact, the answers are closed but not the questions.

The 12 question types presented in Sect. 9.2 could be further discussed as questions that can be presented as story questions and/or as closed questions. In what follows, we classify them according to this criterion.

Types of questions that can be presented as closed questions:
- Type3—Tracing a given solution
- Type4—Analysis of code execution
- Type5—Finding the purpose of a given solution
- Type6—Examination of the correctness of a given solution
- Type9—Efficiency estimation

Example: A closed question of Type6 can present a list of methods that potentially solve a given task, and ask learners to mark the methods that solve the task correctly.

Types of questions that cannot be presented as closed questions:
- Type1—Development of a solution
- Type2—Development of a solution that uses a given module
- Type10—Question design
- Type12—Transformation of a solution

Since the target of these types of questions is to require learners to develop a solution that meets specific requirements, if they become closed questions, this target will not be achieved.

Types of questions that cannot naturally be presented as closed questions:
- Type7—Completion of a given solution
- Type8—Instruction manipulations
- Type11—Programming style questions

Example: A closed question of Type7 can present a list of optional instructions to be added to a given code in a specific place in order to solve a given task, and learners are asked to mark which of them fits for this purpose.

The target of Activity 77 is to train the prospective CS teachers in question formulation. This training is especially important in the context of story questions which usually include more words.

In ITiCSE-13, a working group developed a set of 654 multiple-choice questions (a kind of closed question) on CS1 and CS2 topics called the Canterbury Question-Bank (Sanders et al. 2013). This work describes twelve patterns of multiple-choice questions, labels each one with a name, and adds a short description, a sample question(s), and an identification of which of the twelve types of questions, presented in this chapter, fits it the most (Sects. 9.2.1–9.2.12). The QuestionBank is publicly available as a repository for computing education instructors and researchers.

Activity 77: Question Formulation, Work in Pairs
First, the students work in pairs and each mate of the pair is asked to develop an open question on one specific topic from the curriculum. Then, they are asked to exchange the questions they write, solve the question that their mate develop, and give a constructive feedback about the question to their mate. It is important that both mates will develop a question on the same CS topic, since during the question development they deepen their understanding and consideration with respect to the specific topic, and consequently, their feedback to their mate is more valuable.

9.4.3   Unsolvable Questions

The rational for this specific short discussion stems from the fact that CS learners should be aware of the fact that not every problem is solvable (see also Chap. 5). There are incomputable problems—problems that have no solution, meaning, there is no algorithm that solves them, and further, a proof exists that shows that such algorithm does not exist. In addition, there are problems that have an algorithmic solution, but they cannot be computed in practice due to their time complexity.

It is important that a CS educator be aware to the fact that sometimes there is no hint at whether a problem is solvable or not. Further, from a high school CS teacher's point of view, the message that should be delivered is that any question design should be done very carefully. Sometimes questions may look simple, but their solution may be very complicated or even does not exist.

Hazzan (2001) addresses this kind of question by focusing on their presentation to CS learners. The idea is to formulate a problem in a way that would not give any hint of whether the problem is solvable or not, or of the conditions under which it is solvable. Such a formulation has two main merits. First, learners get the idea that there are unsolvable problems, and second, learners acquire skills for determining whether a problem is solvable or not, and if a problem is solvable under certain conditions, to find out these conditions. This idea is illustrated in Hazzan (2001) with respect to different problems, for example, the halting problem, the tiling problem, and map coloring problem.

9.5   Assimilation of the Types of Questions to Different Computer Science Contents

As we stated in the beginning of the chapter, the presented types of questions are mainly programming-related questions. Still, most of them can be used in a variety of CS contexts. Table 9.4 presents specific variations of several question types in the context of Automata theory.

Table 9.4 Illustration of the types of questions in the context of Automata theory

Type of question	An example pattern
Type1: Development of a solution	Design a finite automaton that recognizes a regular language L
Type2: Development of a solution that uses a given module	Given the A1 finite automaton that recognizes the language L1 and A2 finite automaton that recognizes the language L2, design a finite automaton that recognizes the language L1 ∪ L2
Type3: Tracing a given solution	Given a push down automaton P, and a word w, show the sequence of states that P goes through while processing w
Type4: Analysis of code execution	Given a finite automaton A, present A word that the automaton accepts A word that the automaton rejects A word that its processing is terminated in the trap state
Type5: Finding the purpose of a given solution	Given a Turing Machine T, determine what language it accepts
Type6: Examination of the correctness of a given solution	Does the given Turing Machine T recognize the language L?
Type7: Completion a given solution	Complete the Push Down Automaton P, so it will recognize the language L
Type8: Instruction manipulations	Given a Turing Machine T, if the transition from state q1 to q2 is replaced by the next transition [to be described], what language will the machine recognize?
Type9: "Efficiency" estimation	Given a finite automaton A that recognizes the language L, can you present a different finite automaton that recognizes the same language with fewer states?
Type10: Question design	Design a question that requires the presentation of a BNF grammar for an irregular language
Type11: "Programming" style questions	Given three different Push Down Automata that recognize a language L, examine the automata and state which of them, in your opinion, is the best. Explain your answer
Type12: Transformation of a solution	Given a Turing machine T, present a BNF grammar that expands the same language

The construction of such a table can be facilitated as an activity in the MTCS course for one of the topics included in the high school curriculum that the MTCS focuses on.

9.6 Question Preparation

The preparation of questions to be used in a CS class is not a simple task. Many considerations should be thought about, some of them are local to the specific lesson and class and others are more global and refer to the teaching unit, or even, further, to the entire curriculum. Those considerations are important and therefore, the process of question preparation should be included in CS teacher preparation programs. Though an entire course can be dedicated to aspects related to question preparation in different stages of the teaching processes, we present here only the main stages of the preparation process of a question to be used in a CS class.

1. *Planning*: We lay out questions that CS teachers should ask and answer in the process of question planning:

 - What is the target of the question?
 - What does the question intend to examine?
 - What knowledge and skills are students supposed to possess in order to solve the question?
 - Does the previous learning process enable students to acquire those skills?
 - Does the previous learning process include the needed knowledge?
 - What is the level of abstraction needed to solve the problem?
 - Is the question varied from other questions presented so far in the class?

2. *Solving*: Teachers must solve any question before presenting it to the learners. It is important for any question, but especially necessary in the process of test preparation (see Chap. 10). Indeed, there are questions that look simple, but turn out to be difficult. Until a complete solution is presented, one cannot be sure that the question fits its purpose as well as the other aspects presented in the planning stage. If possible, it is recommended to ask a colleague to read at least the question formulation and verify that the question is understandable, clear, and not ambiguous. When the answer is based on code writing, it is necessary to run it on the computer and check that it works properly.

3. *Estimation of the needed time to solve the question*: Time is a crucial resource in teaching processes. Therefore, teacher should estimate the time required by learners to solve any particular question, which is usually longer than the time required by the teacher to solve it. The time estimation in this context relies on different factors, such as the effort involved in reading, planning, and writing the answer to the question. If a specific question turns out to be time consuming, it is important to remember that, in most cases, a different type of question, that requires less effort and meets the same pedagogical targets, can be developed.

The target of Activities 78 and 79 is to train the students in question design of different types. Additional activities about evaluation in general and about test preparation in particular are presented in Chap. 10.

Activity 78: Question Design, Individual Work or Group Work
All students/groups are directed to focus on one specific CS subject, such as, the loop or differences between indices and values of array cells. In addition, for each student/group, one type of question is assigned. Then, each student/group composes a question of that specific type related to the specified CS subject. The students can exchange the questions to get feedback, and based on it to improve the questions formulation.

In the next stage, all questions are merged into one document. Such a document highlights very clearly how a specific subject can be addressed by a variety of types of questions.

Further, the course instructor can facilitate a discussion about how to evaluate question difficulty, by posing questions such as: Can we definitely decide whether a question is simpler or more difficult than another question? What elements do determine question difficulty? Can we sort questions according to their difficulty level? Is question difficulty level connected directly to its type?

This discussion can follow by an activity in which for several of the presented questions, the students are asked to formulate two similar questions—one that is simpler and one that is harder. An interesting question that the students can be asked to reflect on is whether different levels of question difficulty require also different question types.

Activity 79: Test Design, Group Work
This activity can be facilitated in the context of any topic discussed in the course. The activity is presented in Table 9.5.

The students are divided into groups and are asked to compose a test on a specified CS subject taken from the high school curriculum that the MTCS course focuses on. Each question in the test should represent a different question type. While working on this activity, the student should relate to the different stages of question composition, to the variety of question types, and to the CS contents as well.

Table 9.5 Worksheet on test design

Worksheet—test design, group work
Compose a test on (conditions/loops/arrays/…) according to the following stages:
1. Determine which concepts should be tested
2. Decide what types of questions to include in the test. Associate the concepts to be checked to each type of question and roughly estimate the time required for a high school pupil to solve it
3. Distribute the work among the team members, so that each team member will compose one question
4. When all team members finish composing their questions, organize them into one test
5. Re-estimate the time required for a high school pupil to solve each question and check if the total time estimations fits the time framework of the test
6. Each team member solves the test and writes down his or her comments on each question
7. In your groups, discuss your comments and perform the needed changes in the test
8. Submit the test to the course web site

References

Biggs JB, Collis K (1982) Evaluating the quality of learning: the SOLO taxonomy. Academic Press, New York

Bloom BS, Engelhart MD, Furst EJ et al (1956) Taxonomy of educational objectives Handbook 1: cognitive domain. Longman Group Ltd, London

Boyer KE, Lahti W, Phillips R, Wallis MD, Vouk MA, Lester JC (2010) Principles of asking effective questions during student problem solving. Proceedings of the 41st ACM Technical Symposium on Computer Science Education (SIGCSE '10) ACM, New York, USA, pp 460–464

Hazzan O (2001) On the presentation of computer science problems. Inroads—the SIGCSE Bull 33(4):55–58

Jones KO, Harland J, Reid JM et al (2009) Relationship between examination questions and Bloom's taxonomy. In Proceedings of the 39th IEEE International Conference on Frontiers in Education Conference San Antonio, Texas, USA

Pomerantz J (2002) Question types in digital reference: an evaluation of question taxonomies. In Proceedings of the 2nd ACM/IEEE-CS Joint Conference on Digital Library. Portland, Oregon, USA

Ragonis N, Shilo G. (2013) What is it we are asking: Interpreting problem-solving questions in computer science and linguistics. In Proc. of the 44th ACM Technical Symposium on Computer Science Education (SIGCSE '13). ACM, New York, USA, pp 189–194

Sanders K, Ahmadzadeh M, Clear T, Edwards SH, Goldweber M, Johnson C, Lister R, McCartney R, Patitsas E, Spacco J (2013) The canterbury questionbank: building a repository of multiple-choice CS1 and CS2 questions. In Proceedings of the ITiCSE working group report Conference on Innovation and Technical in Computer Science Education -working group reports (ITiCSE -WGR '13). ACM, New York, USA, pp 33–52

Thompson E, Luxton-R A, Whalley J et al (2008) Bloom's taxonomy for CS assessment. In Simon, Hamilton, M (Eds), Proc 10th Australas Comput Educ Conf (ACE 2008), vol 78. Wollongong, NSW, Australia. CRPIT, pp 155–162

Yang A., Wu J, Wang L (2008) Research and design of test question database management system based on the three-tier structure. WTOS 7(12):1473–1483

Assessment 10

Abstract

Assessment is one of the most common tasks teachers perform from the early stages of their professional development. This chapter highlights the uniqueness of learners' assessment in the case of computer science education, emphasizing that assessment is not a target by itself, but rather, a pedagogical means by which (a) teachers improve their understanding of the current knowledge of their learners and (b) learners get feedback related to their own understanding of the learned subjects. The chapter also delivers the message that the theme of assessment can be discussed in the Methods of Teaching Computer Science (MTCS) course in different opportunities, for example, learners' alternative conceptions, project-based learning, and types of questions. This chapter focuses on tests, peer assessment, project evaluation, and the use of portfolio in computer science education. We end this chapter by addressing the assessment of the students enrolled in the MTCS course.

10.1 Introduction

Assessment is one of the most common tasks teachers perform from the early stages of their professional development. Therefore, and in order to highlight the uniqueness of learners' assessment in the case of computer science (CS) education, the topic should be included and addressed in the Methods of Teaching Computer Science (MTCS) course.

Assessment and grading are not the same. The goal of grading is usually to evaluate individual knowledge and performance; nevertheless, grading does not always measure reliably the learning process and the learner improvement. The goal of assessment is to improve student learning; it can include many ungraded measures of student understanding and can be used also to improve educational practices.

© Springer-Verlag London Limited 2014
O. Hazzan et al., *Guide to Teaching Computer Science*,
DOI 10.1007/978-1-4471-6630-6_10

The constructivist teaching-learning approach should be reflected also in the assessment process, which, in order to be meaningful, can be tightly intertwined in the teaching and learning processes. Thus, assessment is carried out as part of the teaching process, for example, by observing learners' engagement in the learning process in general, and in particular, during class discussion and problem-solving situations. Such situations provide the teacher with an opportunity to explore learners' thinking processes during the construction process of their knowledge.

Therefore, it is important to realize that assessment is not a target by itself; but rather, it is a pedagogical means by which: (a) teachers improve their understanding of the current knowledge of their learners, and (b) learners get feedback related to their own understanding of the learned topic. Accordingly, the main target of assessment should not be grading; alternatively, assessment should serve as a kind of reflection both for teachers and learners with respect to the teaching and learning processes and with respect to learners' knowledge and perceptions.

Educational practices distinguish between formative assessment to summative assessment.

Formative Assessment Formative assessment should be regularly integrated in the teaching-learning process. Its target is to monitor student learning and to provide ongoing feedback that can be utilized by instructors to improve their teaching and by students to improve their learning. Formative assessment helps learners identify their strengths and weaknesses, and at the same time help educators recognize learners' difficulties and address them in the teaching-learning process. In the CS class, formative assessment can be done regularly, for example, during lab exercises or by development of small projects. It can be carried out also by presenting students specific tasks for this goal, such as, to draw a concept map to represent their understanding of a topic or to submit one or two sentences which encapsulates the main ideas presented in a lesson.

Summative Assessment The target of summative assessment is to evaluate learner knowledge and skills at the end of some learning process, such as the end of the instruction of a specific unit by addressing the instructor learning targets, or at the end of a project development. Summative assessment is usually used for grading, and it enables to compare learners' achievements. Summative assessment methods include exams, final evaluation of a project, and program checking by automatic testing. Summative assessments can be then used for formative assessment when learners or educators use it to guide their efforts and perform some subsequent activities in the class course of study. For example, if a class fails in a specific question in a test, it could be either because the question was not formulated clearly, and the educator can learn about future formulation, or because the learners did not understand some concepts, and the educator can reflect and teach these concepts again perhaps by using different teaching-learning tools.

Two more assessment tools that can be used both as formative and summative assessment are self-assessment and peer-assessment.

Self-Assessment In self-assessment students assess their own work. In the last years, as part of conceiving assessment and reflection as learning tools, self-assessment is commonly integrated in learning scenarios. Self-assessment is applied by CS studies as well. For example, based on Bloom's revised taxonomy, Alaoutinen (2010) evaluated a new taxonomy for self-assessment scale, and examined factors that affect assessment accuracy and course performance. The study showed that students can locate their level of knowledge (i.e., to assess) well along the taxonomy-based scale and that the scale fits well engineering students' learning style. The study pointed that advanced students assess themselves more accurately than novices, and that reflective students are better in programming. Those results can encourage educator to use self-assessment as fitting both for reflection and improvement of students' knowledge and professional skills.

Peer-Assessment In peer-assessment students' work is assessed by other students of equal status. Peer-assessment can be done in conjunction with self-assessment. Students reflect on their own efforts, and extend and enrich this reflection by comparing their own feedback with their peers' assessment. Peer assessment is a powerful meta-cognitive tool, it engages students in the learning process and develops their ability to reflect on and critically evaluate their learning and skills development (Li 2011, collection of related publication[1]). In CS education, peer-assessment in the classroom is explored by Turner et al. (2011), and Sitthiworachart and Joy (2004) discuss benefits of peer assessment in the context of undergraduate courses.

Similar to the use of other pedagogical tools, the assessment methods employed in any pedagogical setting should be varied. In this spirit, we deliver the message that learners' assessment should not be necessarily addressed in isolation in the MTCS course and recommend to integrate activities and discussions about assessment along the MTCS course in different opportunities when other topics are at the focus of the discussion, for example, research in CS education (Chap. 4), learners' alternative conceptions (Chap. 6), lab-based teaching (Chap. 8), types of questions (Chap. 9), and teaching planning (Chap. 11).

For example, Activity 45 presented in Sect. 6.3 is about learners' alterative conceptions as well as about the assessment of a learner's answer in a written exam; in Chap. 7, we discuss project-based learning and mentoring process of software project development. The theme of assessment can be discussed with respect to different approaches towards project evaluation. Finally, we mention Chap. 9 that presented types of questions, which also elevate relevant themes related to learners' assessment.

The following topics are addressed in the chapter: tests, project assessment, and portfolio in CS education. We end by addressing the assessment of students enrolled in the MTCS course. These topics enable to deliver the following principles of assessment in the context of CS education:

---

[1] Melbourne university, student peer review. http://peerreview.cis.unimelb.edu.au/resources3/publications/

- A single and unique way to assess CS learners does not exist and different assessment approaches are appropriate to be applied in different pedagogical situations.
- Different aspects of learner's knowledge and cognitive skills should be assessed.
- Assessment tasks should be varied in order (a) to relate to different aspects of the learned topic and different cognitive skills and (b) to motivate learners and keep their curiosity.
- Assessment should be conceived as an ongoing reflective process.
- The different assessment approaches should make sense, and when appropriate, should be explained to the learners' (who take the exam, develop the project, etc.).
- Teachers' feedback to learners' exercises/exams/projects may convey different messages (sometimes hidden); therefore, careful attention should be given to written feedbacks.

Formative assessment serves best learners knowledge construction. Our recommendation is that formative assessment will be integrated along the entire learning process using different class activities as presented in different chapters of the book. In particular, formative assessment can be integrated in the spirit of active learning (Chap. 2), while emphasizing problem-solving approaches (Chap. 5) and in lab-based assignments (Chap. 8).

Activity 80: Practicing Self-Assessment and Peer-Assessment
Clearly, this practice can be used in any CS course and in particular in the MTCS course. In the later, it can be facilitated at any time when students are asked to submit a self learning outcome.

- Stage A: Ask students to perform the said task as an individual work.
- Stage B: Ask each student to: (1) evaluate her/his outcome; (2) reflect on the self-assessment process; (3) present the measures on which the evaluation was based.
- Stage C: Students exchange their outcomes and evaluate a peer's work as an individual work.
- Stage D: Each peer compares and discusses the measures each one used for the evaluation. This is done as a peer work.
- Stage E: Facilitate a class discussion on the affective and effective aspects of both the self-assessment and the peer-assessment. Guide the discussion to highlight the following main aspects: in order to meaningfully perform self or peer assessment, a prediscussion on the actual measures should be conducted; guidelines or a rubric should be provided as a base for these kinds of assessment; self-assessment is an effective task when reflective processes are inherited in the class learning processes; peer-assessment is mainly effective when pleasant and constructive class climate is established.

In what follows, we relate to tools that can be seen as summative assessment tools, and focus on how to use them to achieve learning targets as well. We do not discuss the automated programming assessment since it does not address the quality of programming and programming style.

10.2 Tests

Tests can be administered in different forms and settings. For example, tests can take place in the computer lab. In this section, however, we focus on written tests which take place in the classroom. This process of test handling is based on several steps:

1. The teacher constructs the test and the test evaluation rubric.
2. Students take the exam.
3. The teacher evaluates the test.
4. The teacher returns the tests to learners.
   Since steps 2 and 4 are more general and basically, are similar to those performed in other subject matters, Activities 81 and 82, focus on steps 1 and 3.

Activity 81: Test Construction

It is recommended to facilitate this activity after different types of questions in CS education (Chap. 9) have been discussed. The instructor of the MTCS course can choose any CS topic to focus on in this activity according to his or her pedagogical preferences, the topics addressed so far in the MTCS course, and the relevant high school curriculum.

- Stage A: First experience in test construction, individual work
  In this trigger, the students are asked to construct a test about a CS topic. The instructor can specify the stage in which the test takes place (that is, during the learning process of the said topic, at the end of this learning process, etc.) or to leave it open for the students to decide. The idea is to let the students start sensing the variety of topics that a teacher should consider while building a test. Therefore, the students should not necessarily finish the task and the instructor can proceed to the next stage when he or she recognizes that the students have gained enough experience that will enable them to discuss meaningfully the process of test construction.
- Stage B: Topics to be considered in the process of test construction, class discussion
  Based on the experience the students have gained in Stage A, a class discussion takes place about the topics that a teacher should address when he or she constructs a test. The instructor should make sure that the following topics are mentioned in this discussion: the targets of the test, the structure of the test, learners level, types of questions, questions of different complexity levels, organization of the questions in the test, and the grading

policy. This discussion aims to increase the prospective teachers' awareness to the fact that the process of test construction is not a trivial task and that special attention should be devoted to a variety of pedagogical considerations.

- Stage C: Test construction, group work
  At this stage, after the students have realized the variety of topics that should be addressed while a test is constructed, they are asked to work in groups and to construct a test on the topic they worked on in Stage A.

First, the students are asked to sketch the structure of a test. They should be guided to decide about the questions' scope (in terms of their relative grade in the test) and type, and to document their pedagogical considerations and decision-making processes.

For example, with respect to question scope, the test can include either many short and focused questions, questions of different scopes, or a small number of wide scope questions. With respect to question type, the students should decide whether the questions are open or closed, programming tasks, etc. It should be emphasized that the decided framework can be changed later when the details of the test are examined; however, it should be highlighted that such an initial sketch of the test structure helps teachers clarify to themselves their pedagogical purposes and what kind of knowledge they actually wish to evaluate.

Second, the students are asked to develop the questions of the test according to the structure of the test they decided about. During this process, if they decide to change that structure, they should explain why.

The process of question construction encourages the students to consider topics such as, factors which determine the complexity level of a particular question, different solutions that learner may propose for a given question, learners' mental processes, and the need to match questions to a specific group of learners.

- Stage D: Test analysis, class discussion or teamwork
  One way to facilitate this stage is to let each group present its test together with its pedagogical considerations.

Another way to facilitate this stage, especially when the class is big and the available time does not allow all the groups to present their tests, is to implement peer review and let each group analyze the test of another group (e.g., the test of the group on its right), comment about it and returns it to the group that designed it. Thus, in parallel, all groups analyze a test that was constructed by another group, and immediately after that, receive another group's feedback on the test they constructed. This teamwork can proceed by short presentations in which each group presents the test it analyzed.

Then, one test is selected on which all groups will work during the next stages. It is recommended to let the students select the test on which they will work. However, if the students do not reach a conclusion, the instructor of the MTCS course can choose one test, explaining his or her considerations. For example, the selection can be based on the variety of questions, reasonable amount of material covered by the test, and so on. Alternatively, the test on which the students will work can be formed by integrating questions taken from tests presented by different groups.

- Stage E: Test solving, individual work
  The students are asked to work individually and to solve the selected test. This process of test solving aims at increasing the students' awareness to a variety of topics (such as, the relevance of the questions to what was learned in class, the solution complexity, the scope of the solution, learners' potential mistakes, and more) that they will have to consider in the next stage in which they construct an evaluation rubric for the test.
- Stage F: Design of an evaluation rubric, group work
  An evaluation rubric is a set of guidelines that a teacher uses in the grading process of a specific test. The actual preparation of evaluation rubrics encourages teachers (a) to realize what the test actually checks and if it matches their pedagogical intentions, (b) to verify that there is a match between the grades the learners will get and their actual knowledge, and (c) to ensure (as much as it is possible) that all learners' exams are checked uniformly by the same criteria.

In addition, teachers can share the evaluation rubric with their pupils, when they wish to deliver what they consider important with respect to the test content or when they wish to explain to their pupils how their grades were calculated. When the evaluation rubric is shared with the pupils, it is not necessary to indicate all its details; rather, each teacher should select the level of detail he or she shares with his or her pupils according to the class characteristics and his or her personal pedagogical considerations. In any case, it should be delivered to the students that it is important for their future pupils to be familiar with their evaluation principles.

When an evaluation rubric is designed, a teacher should consider several aspects:

- *Point Accumulation* Should points be gathered (i.e., a pupil starts with zero points and collects points according to his or her answers) or should points be reduced (i.e., a pupil starts with 100 points, and mistakes reduce his or her grade)? Each approach is appropriate as long as it is based on relevant pedagogical considerations.
- *What Is Considered a Mistake* For example, if a pupil wrote a correct computer program but did not use meaningful names for methods, should

points be reduced (if the approach is to subtract points)? If a student found a solution to a given problem, described it correctly, but did not implement it correctly in the programming language, should points be added (if the approach is to add points)?

- *Evaluation of Different Solutions* If a question can be solved in several ways, is one answer preferable over the others? Are all the solutions accepted?

  After these considerations are discussed with the students, they are asked to work in groups and to construct an evaluation rubric for the test selected in Stage D which they solved in Stage E. They are asked also to choose one team member to document their considerations and decision-making processes.

- Stage G: Presentation of the evaluation rubrics

  The groups present their evaluation rubrics in front of the class, explaining their considerations with respect to each question. In order to highlight the fact that the same question can be evaluated in different ways, it is recommended to ask all groups to present their evaluation rubrics for the first question, then for the second question, etc. If there are time constraints, the instructor can choose one or two questions and to focus on their evaluation rubrics.

As in Stage D, another option is to let each group analyze the evaluation rubric of another group and to present its conclusions in front of the class.

- Stage H: Conclusion

  The instructor of the MTCS course summarizes the main issues addressed in the different stages of this activity.

Activity 82: Construction of a Course-Summary Exam
In this activity, a course-summary exam (e.g., the AP exam) is constructed. In some sense, this activity is similar to the previous activity (Activity 81); it is, however, carried out with respect to a different scope of CS content and learner population. From the content perspective, a course-summary exam evaluates learners' knowledge with respect to all the subjects included in the course; from the learner population perspective, a course-summary exam is not intended to be solved by one (or small number of) specific class, but rather by the entire course population. Definitely, these larger scales set a challenge.

In addition to the importance of letting the prospective CS teachers experience test design and construction processes, we highlight two additional pedagogical advantages of the facilitation of this task in the MTCS course. First, in order to develop a course-summary exam, the prospective CS teachers should review the entire course curriculum, and as a result, it is reasonable to assume that they deepen their familiarity with this curriculum. In countries, where formal summative tests exist, this activity is an appropriate opportunity to explore previous formal tests as part of the preparation of the prospective teachers for their field work at schools. Second, while building the exam questions, they should consider the notion of diversity (see Chap. 3) in order to adopt the exam to a wider learner population that will take it.

Specifically, in this activity, the students work in pairs and construct a course-summary exam together with its evaluation rubric. Then, the following stages can take place:

1. Each pair exchanges its test with another pair and each student solves individually the test that the pair received.
2. Each pair checks the exams of the two students with which they switched the exams according to the valuation rubric it prepared.
3. Based on the exam evaluation, if needed, each pair updates the exam and its evaluation rubric.

10.3 Project Assessment

One of the topics discussed in Chap. 8, which deals with lab-based teaching, is the integration of software projects in CS education. Here, we explain the rationale for project-based learning and concentrated on the actual mentoring process of software projects developed by CS learners. Project evaluation is an additional important issue that should be addressed in this context. Specifically, questions such as the following ones should be considered: How should projects be evaluated? What should be the nature of the evaluation? When should the evaluation take place? How should software projects developed by teamwork be evaluated? How can project evaluation enhance learners' understanding of CS?

Project evaluation is not a simple task and therefore, should be addressed in the MTCS course. We mention that the focus of this chapter will be placed on the evaluation of software projects. Clearly, nonsoftware projects can also support learning processes of CS. See, for example, Activity 97 in Sect. 12.7 for a discussion about the evaluation of nonsoftware projects.

In what follows, we first present several approaches for project evaluation. Then, we suggest several activities related to project evaluation to be facilitated in the MTCS course.

We address the evaluation of two kinds of projects: software projects developed by individuals and software projects developed by teams.

10.3.1 Individual Projects

Meerbaum–Salant and Hazzan (2010) suggest three resources for the evaluation of software project developed by pupils individually: the teacher, peers (i.e., other pupils in the class), and the learner who develops the project. We elaborate on each of them.

- *Teacher evaluation* can be performed in two ways:
  - *Formative Project Assessment* This assessment is carried out by the teacher during the entire process of project development with respect to (almost) each activity that the student performs. The purpose of formative assessment is to guide the pupils in the development process in order to support their development process and improve their understanding of the relevant CS contents.
  - *Summative Project Assessment* The teacher can perform summative assessment several times during the development process, usually at the end of specific stages, to monitor the students' and class' progress.
- *Peer project assessment* can be carried out, for example, in the following way: The pupils are divided into groups. Each pupil presents his or her project to the other group members and receives their feedback.
- *Individual project feedback/evaluation* can be encouraged by asking each pupil to reflect on his or her work and on the way he or she plans to meet the schedule that the teacher set for the entire class. See Chap. 5 for a broader discussion about reflection and reflective processes.

10.3.2 Team Projects

Software projects developed by teams are common in undergraduate CS education, and specifically, in capstone courses that the students study in their senior year. This attention is reflected, for example, in the SIGCSE 2014 conferences: 15 out of 105 presented papers referred to the integration of software projects in different academic courses (e.g., Brown et al. 2014) as well as to their evaluation (e.g., Vasilevskaya et al. 2014). In these courses, undergraduate CS students develop a software project, in most cases in teams, that encapsulates what the students have studied during their undergraduate studies.

Studies that address student software projects usually deal with issues such as the assignment of students to groups (Redmond 2001; Smith and Smarkusky 2005; Bender et al. 2012), the coordination of teamwork (Moses et al. 2000), the grading of such projects (Chamillard and Merkle 2002), and ways by which instructors can gain information about the contribution of individual students to the team project (Lawhead and Wilkins 2000; Vasilevskaya et al. 2014).

In this spirit, the following discussion about the evaluation of projects developed by teams is especially relevant for software projects developed by undergraduate students, but nevertheless can be applied also in the high school setting.

According to Hazzan (2003), the evaluation of software projects developed by teams is analogous to reward allocation to software teams in the industry. The topic of reward allocation with respect to the profession of software engineering is important for several reasons. We mention three reasons which are also relevant for the evaluation of software projects developed in educational frameworks:

- Teamwork is essential for software development. As a result, conflicts between the contribution to the teamwork and the way by which rewards are shared may intensify.
- Software developers are usually highly motivated. This can cause conflicts between personal targets and team goals.
- Team-based rewards may cause social problems, such as the free-rider phenomenon.

Accordingly, a relevant question addressed by the literature is whether to distribute incentives among team members equally or not.

Dubinsky and Hazzan (2005) suggest a grading policy for software projects developed by teams of undergraduate CS students which aims at motivating both teamwork and collaboration as well as the personal contribution of each team member to the project success (see Table 10.1). The grading policy is composed of two main components. The first one is a group component (65 %) whose main criterion is the meeting of the customer stories as well as the time estimations given by the students at each of the three iterations in which the projects were developed throughout the semester. The second ingredient of the grading policy is an individual component (35 %), whose main criterion is the personal performance of the student with respect to his or her development tasks as well as with respect to his or her personal role in the project.

Table 10.1 A grading policy for software projects developed by teams. (Dubinsky and Hazzan 2005)

Group component (65 %)	Individual component (35 %)
60 %—answer the customer stories and meeting the schedule according to the team time estimations:	50 %—weekly reflection Pair programming experience
(10 %) for iteration 1	Test-Driven-Development exercise
(25 %) for iteration 2	Weekly presence
(25 %) for iteration 3	25 %—performance of a personal role:
25 %—project documentation	Actual implementation
15 %—group evaluation of the academic coach	25 %—Further development and enhancement personal evaluation of the coach

Activities 83–85 address project evaluation and aim to further increase students' awareness to the challenges involved in this process.

Activity 83: Getting Familiarity with an Evaluation Rubric for Software Projects
An assessment rubric can serve project evaluation either by a group or as self-assessment (Vivar et al. 2013). In this activity, the students in the MTCS course are presented with the main categories of an evaluation rubric for software projects which was developed by two high school CS teachers independently (see Table 10.2). The two teachers teach two 11th grade classes in parallel, and since each teacher examines the other teacher's class, he or she is not familiar with the actual details of the projects he or she evaluates.

The projects for which this evaluation rubric was constructed are developed by the pupils individually, for about half a school year, when each week the pupils dedicate about 3 hours for the project process. The material needed for the development process is learned prior to the development process.

The grade is determined based on an examination of the project file (which includes the project documentation, its code, and its scope) and an oral exam. The oral exam takes place in the computer lab, in which, in order to observe the pupil's familiarity with the code and its functionality, the teacher asks the pupils specific questions about the project, as well as to modify the project in specific ways and to add code with specific functionality.

The students in the MTCS course are asked to work in pairs and:

1. To identify the pedagogical purposes of the teachers who developed this rubric.
2. To specify subcategories of the main categories of the evaluation rubric.
3. To suggest how they would change the evaluation rubric, if at all, to fit it to their pedagogical approach.

Table 10.2 Example of an evaluation rubric for software projects developed by high school pupils individually

Topic	Max points	Actual grade	Comments
Project documentation and organization	10	–	–
Project code	10	–	–
Project scope	30	–	–
Knowledge about the project and its domain	30	–	–
Extension and changes during the lab exam	20	–	–
Total grade	100	–	–

Activity 84: Construction of an Evaluation Rubric for Software Projects

- Stage A: Setting the teaching scene
  The instructor of the MTCS course selects a project theme and scope to be developed by high school pupils and describes it to the students of the MTCS course. One such setting is presented in Activity 83. Here are two additional examples of such descriptions:

Example 1: 11th grade pupils develop a software project in the object-oriented development paradigm approach during the entire school year. They already learned the basic CS concepts in the 10th grade. If the project requires additional knowledge, the pupils learn this knowledge by themselves with their teacher' aid. Each week they should dedicate 6 hours to the project development.

Example 2: 10th grade pupils develop a software project in pairs for 2 month in the Alice development environment (see Chap. 8). They learn the CS material in parallel to the project development. Each week they should allocate 4 hours to the project development.

After such a description is presented to the students, the instructor asks the MTCS course students what aspects of the project, in their opinion, should be evaluated. Their suggestions are listed on the board.

- Stage B: Construction of an evaluation rubric, group work
  The students are asked to work in groups and to construct an evaluation rubric for the project described in Stage A. They are also asked to document their pedagogical considerations for each decision they took in that construction process.

After the students have developed the evaluation rubrics, each group presents its rubric alongside its pedagogical considerations.

- Stage C: Conclusion
  In this conclusion, the instructor should highlight the following messages related to evaluation processes in general and emphasize their application and relevance for the evaluation of software projects:
- The purpose of evaluation is to enhance learners' learning processes and understanding.
- Learners should get the message that evaluation is a means rather than a target by itself.
- An evaluation policy should match the educational messages that a teacher delivers to his or her pupils.
- The pupils should be familiar with their evaluation process from the very beginning of the evaluation process.
- Evaluation policies should address both the learning process and its final product.
- Different aspects of the learning process should be assessed.

Activity 85: Analysis of a Grading Policy of the Group Project
The grading policy presented in Table 10.1 for grading software projects developed by teams is presented to the students in the MTCS course. They are asked to work in small teams and:

• To analyze its advantages and disadvantages.
• To describe how it may influence team members' behavior and collaboration.
• To explain how it enables to achieve both group and personal interests.
Such a discussion raises the prospective CS teachers' awareness to the multifaceted nature of the development process of software projects by teams, and to the different considerations they should address in the evaluation process of such projects.

10.4 Portfolio

A portfolio is a collection of learners' works which reflects learners' progress and achievements in a specific domain along the learning process. Since a portfolio is prepared by the learners along a period of time, it can be viewed as a formative assessment tool (see Sect. 10.3) and it is important to consider its position and role when a teaching process is planned (see Chap. 11).

In more detail, a portfolio is a purposeful collection of student's works that tells the story about learner's efforts, progress, achievement, and self-reflection on his or her learning process and progress, in one or more knowledge areas. The portfolio content should be selected and decided upon together with the learner and, as with respect to other evaluation tools, its evaluation criteria should be clear to the learners from the early stages of the portfolio construction process (Arter and Spandel 1992).

This description implies that the portfolio items should not be selected randomly, but rather, that they should be carefully chosen together by the teacher and the pupil in a way that indicates that learning occurred, represents the learner's achievements and progress, and reflects the learner's knowledge and skills with respect to specific domains. In addition to learners' products and other teachers' evaluation tools of the learner's learning process, a portfolio can include teachers' observation during the learning process, peer reviews, and the learners' suggestions for how to continue their learning process of the said topic. Patton and McGill (2006) reported on the use of portfolio in a CS course. They mentioned the pedagogical advantages of using portfolio, among them the ability to conduct a longitudinal study of student performance, course assessment, and the prevention of cheating.

These characteristics of the portfolio make it a pedagogical tool that:

• Integrates learning with assessment
• Creates a continuous communication and collaboration channel between teachers and learners

Table 10.3 Professional tips—planning and using portfolio assessment. (Hayes 1998)

Tip #1:	Develop a portfolio assessment process specific to your own situation
Tip #2:	Use a collaborative planning approach that involves the teachers and learners who will use the portfolio process
Tip #3:	Define the purposes and audiences for portfolio assessment PRIOR to developing the portfolio materials
Tip #4:	Implement the process gradually, allowing time for experimentation and improvement
Tip #5:	Recognize the importance of students as partners in the assessment process
Tip #6:	Plan additional time for students to construct portfolios, student–teacher conferences, and teacher review of portfolios
Tip #7:	Plan outgoing opportunities for staff development to address new skills required for portfolio assessment

- Provides a comprehensive view of learners' achievements with respect to a variety of concepts
- Enables learners to identify both their weaknesses and their strengths
- Encourages learners to take responsibility on their learning process
- Enhances learners' reflective skills

With respect to portfolio assessment, Hayes (1998) claims that it is a complex process that requires considerable planning and decision-making process, and offers seven tips to keep in mind in the process of portfolio assessment (see Table 10.3). It is worth noticing Tip #7, which specifically indicates teachers' need to develop new assessment and instruction skills, such as methods for developing students' reflective skills, or strategies for assessing affective learning outcomes. We suggest that the MTCS course is one venue in which such professional development and training of CS teachers can start, and suggest that one way to evaluate the students enrolled in the MTCS can be based on a portfolio which the students construct during the course of learning (see Sect. 10.5 below).

In the case of CS education, it is relevant to create an online portfolio, which is called ePortfolio: "In general, an ePortfolio is a purposeful collection of information and digital artifacts that demonstrates development or evidences learning outcomes, skills or competencies. The process of producing an ePortfolio (writing, typing, recording, etc.) usually requires the synthesis of ideas, reflection on achievements, self-awareness and forward planning with the potential for educational, developmental or other benefits. Specific types of ePortfolios can be defined in part by their purpose (such as presentation, application, reflection, assessment, and personal development planning), pedagogic design, level of structure (intrinsic or extrinsic), duration (episodic or lifelong) and other factors."[2] We mention that LMSs (Learning Management Systems, such as, moodle) can serve as a platform in which the ePortfolio items are stored and evaluated.

---

[2] Source: http://www.eportfolios.ac.uk/definition

In the case of CS education, a portfolio can include learners' individual and group projects, intermediate versions of these projects, a description of the process in which these projects were developed, peer reviews (see Sect. 10.1), learners' presentation of their work, learners' reflective assessment of their learning process, teacher observations of learners' learning process, and tests.

Activity 86: The Portfolio in Computer Science Education

- Stage A: Portfolio design, team work
  The students are asked to work in teams and to design a portfolio for a high school CS class. Specifically, they are asked to address the following topics and to explain each of the pedagogical decisions:
  - The purpose of the portfolio.
  - The scope of the portfolio, that is, will it focus on a specific CS topic? the learning process during the entire school year? a specific shorter period of time?, etc.
  - The specific period of time during which the pupils will organize their portfolios.
  - A list of items to be included in a pupil's portfolio that enable to achieve the purpose of the portfolio determined above (the rational for the inclusion of any item should be explained).
  - The portfolio organization (online, off-line, blogs, discussion groups, etc.).
  - An assessment framework (a kind of an evaluation rubric) for the portfolio that includes the assessment time line, the teacher-learner discourse mechanism (online, off-line, face-to-face, etc.), and the actual grading policy of the portfolio.
- Stage B: Group presentations of their portfolio
  At the end of the group work, each team presents its portfolio description (scope, elements, periodical schedule, and assessment framework) together with the pedagogical considerations that guided its work.
- Stage C: Class discussion and summary
  After all groups presented their portfolio, a discussion takes place in which the different portfolios suggested by the different groups are discussed and compared: Do they achieve their pedagogical purposes? If so—how? If not—how should they be changed to meet their purposes? Do the different portfolios reflect the same pedagogical approach or different pedagogical approaches? What are the differences between their pedagogical approaches? What are the advantages and disadvantages of each portfolio? etc.

This discussion about the different portfolios sets the basis for a discussion about the uniqueness of the portfolio as an evaluation tool in CS education. This discussion should address (a) the importance of allowing a gradual development process of software projects, (b) the use of online resources in CS education, and (c) the role and use of computational environments for the portfolio organization and management.

The discussion about the use of computational environments for the portfolio organization can take advantage of the fact that the students in the MTCS course are prospective CS teachers with a relatively advanced knowledge both in CS education and computerized tools.

The lesson ends with the instructor's summary of the main ideas related to the portfolio as an evaluation tool in general and in the context of CS in particular.

10.5   The Evaluation of the Students in the MTCS Course

As is indicated in the introduction to this guide, it is recommended not only to talk about CS education but also to implement its pedagogical guidelines in the MTCS course, to let the prospective CS teachers experience the teaching methods presented in the course before becoming high school CS teachers. The same idea is implied for students' evaluation; that is, the evaluation policy applied in the MTCS course should reflect general evaluation principles of high school CS education (as they are discussed, e.g., so far in this chapter). In other words, student evaluation in the MTCS course should not be based only on one kind of pedagogical knowledge, but rather a spectrum of pedagogical skills, activities, and reflection, as well as CS knowledge, all of which reflect different aspects of student achievements in the MTCS course, should be taken into the considerations in student evaluation. In this spirit, a portfolio, for example, can be a suitable evaluation tool for the MTCS course.

Since different CS teacher preparation programs may emphasize different aspects of CS education, we do not specify one specific evaluation scheme for the MTCS course. Rather, in what follows, we suggest a list of components that can be assessed as part of the students' evaluation. Other components, as well as different weights assigned to each component, and different evaluation mechanisms (by the students themselves, by their peers in the course and by the course instructor) are optional for the courses evaluation. In each specific case, however, it is recommended to explain to the students the pedagogical rationale of the determined course grading policy and to publish it in advance.

Possible elements of student assessment in the MTCS course:

- Active participation in the course: This component delivers the importance of practicing different teaching methods as learners, before applying them in high school CS pedagogical situations.

- Portfolio of pedagogical tasks the students developed during the course: This portfolio can include a matriculation or a course-summary exam, the plan of teaching a unit about a specific CS topic, lab-based worksheet, etc.
- Portfolio of nonpedagogical tasks: mini-research, literature review (e.g., of CS education research papers), problem solving in CS (including programming tasks), a poster/presentation about an advanced CS topic, and more.
- A presentation of an advanced CS topic. Such an experience may help them in the future if they need to teach a CS topic that they did not learn as CS students.
- Reflective diary written throughout the semester: This diary should reflect students' conception of CS education in general and of high school CS teaching in particular.
- Peer teaching in the MTCS itself.
- Mentoring activity or practicum in the high school (if needed to be integrated in the course, see Chap. 13).

References

Alaoutinen S (2010) Effects of learning style and student background on self-assessment and course performance. In Proceedings of the 10th Koli Calling International Conference on Computing Education Research (Koli Calling '10). ACM, New York, pp 5–12

Arter J, Spandel V (1992) Using portfolios of student work in instruction and assessment: a NCME instructional module. Educ Meas: Issues and Pract I 1:36–44

Bender L, Walia G, Kambhampaty K, Nygard KE, Nygard TE (2012) Social sensitivity and classroom team projects: an empirical investigation. In Proceedings of the 43rd ACM technical symposium on Computer Science Education (SIGCSE '12). ACM, New York, pp 403–408

Brown C, Pastel R, Seigel M, Wallace C, Ott L (2014) Adding unit test experience to a usability centered project course. In Proceedings of the 45th ACM technical symposium on Computer science education (SIGCSE '14). ACM, New York, pp 259–264

Chamillard AT, Merkle LD (2002) Management challenges in a large introductory computer science course. Proc 33rd SIGCSE Tech Symp on Comput Sci Educ pp. 252–256

Dubinsky Y, Hazzan O (2005) A framework for teaching software development methods. Comput Sci Educ 15(4):275–296

Hayes E (1998) Professional tips for adult and continuing educators—planning and using portfolio assessment, NC Literacy Resource Center, Raleigh, NC., http://muse.widener.edu/~aad0002/portfoliotips.htm. Accessed 14 July 2010

Hazzan O (2003) Computer science students' conception of the relationship between reward (grade) and cooperation. In Proc. 8th Ann Conf on Innov and Technolog in Comput Scie Educ (ITiCSE 2003), Thessaloniki, Greece, pp. 178–182

Lawhead PB, Wilkins DE (2000) Evaluating individuals in team projects. Proc 31st SIGCSE Tech Symp Comput Sci Educ pp. 172–175

Li L (2011) How do students of diverse achievement levels benefit from peer assessment? Int J Scholarship Learn Teach 5:1–16

Meerbaum–Salant O, Hazzan O (2010) An agile constructionist mentoring methodology for software projects in the high school. ACM Trans on Comput Educ—TOCE 9(4) 21:1–29

Moses L, Fincher S, Caristi J (2000) Teams work—panel presentation. Proc 31st SIGCSE Tech Symp Comput Sci Educ pp. 421–422

Patton AL, McGill M (2006) Student portfolios and software quality metrics in computer science education. J Comput Sci Coll 21(4):42–48

Redmond MA (2001) A computer program to aid assignment of student project groups. Proc 32nd SIGCSE Tech Symp Comput Sci Educ pp. 134–138

Sitthiworachart J, Joy M (2004). Effective peer assessment for learning computer programming. In Proceedings of the 9th annual SIGCSE conference on Innovation and technology in computer science education (ITiCSE '04). ACM, New York, pp. 122–126

Smith HH, Smarkusky DL (2005) Competency matrices for peer assessment of individuals in team projects. In Proceedings of the 6th conference on Information technology education (SIGITE '05). ACM, New York, pp. 155–162

Turner S Pérez-Quiñones MA Edwards S Chase J (2011). Student attitudes and motivation for peer review in CS2. In Proceedings of the 42nd ACM technical symposium on Computer science education (SIGCSE '11). ACM, New York, pp. 347–352

Vasilevskaya M, Broman D, Sandahl K (2014). An assessment model for large project courses. In Proceedings of the 45th ACM technical symposium on Computer science education (SIGCSE '14). ACM, New York, pp. 253–258

Vivar AM, Ana Q, Rogado BG, Ramos Gavilán AB, Martín IR, Ascensión M, Esteban R, Zorrila TA, Martín Izard JF (2013) Application of rubric in learning assessment: a proposal of application for engineering students. In Proceedings of the First International Conference on Technological Ecosystem for Enhancing Multiculturality (TEEM '13), Francisco José García-Peñalvo (Ed.). ACM, New York, pp. 441–446

Teaching Planning

11

Abstract

Teaching planning is one of the main pedagogical activities teachers perform. All the tools, ideas, and perspectives presented in the guide can be used and applied in the process of teaching planning, which is, to some extent, independent of the taught discipline. In this chapter, we offer and demonstrate a top-down approach for teaching planning which takes into account a wide range of considerations, and present several activities to be facilitated in the Methods of Teaching Computer Science (MTCS) course for this purpose. The principles of teaching planning presented in this chapter can serve any computer science educator in any teaching framework.

11.1 Introduction

Planning the teaching process is one of the basic practices any teacher performs, and therefore, prospective teachers should acquire the skills needed to accomplish this multilayered task. We assume that the students (i.e. the prospective computer science teachers) have already learned the basic principles of how to plan a detailed lesson layout in one of the general didactics courses. In this chapter, we illustrate how to guide the students to apply this general knowledge in the context of computer science education and further, to take into the consideration the various aspects, teaching tools, and teaching methods presented in the previous chapters of this Guide. Specifically, we suggest a top-down approach for teaching planning (Sect. 11.2), illustrate it with respect to the teaching planning of one-dimensional array (Sect. 11.3), and present several activities to be facilitated with the prospective computer science teachers in the Methods of Teaching Computer Science (MTCS) course (see Sect. 11.4).

© Springer-Verlag London Limited 2014
O. Hazzan et al., *Guide to Teaching Computer Science*,
DOI 10.1007/978-1-4471-6630-6_11

11.2 Top-Down Approach for Teaching Planning

The top-down approach for teaching planning starts with a broad perspective related to the planning of an entire curriculum (e.g., CS1; see Sect. 11.2.1), continues with the planning of one topic from the curriculum (Sect. 11.2.2), and addresses the planning of a specific lesson (Sect. 11.2.3). We also suggest looking at building an understanding of a concept in a spiral gradient manner (Sect. 11.2.4). In all these stages, we refer to a multifaceted consideration that a teacher should be aware of during teaching planning.

11.2.1 Broad Perspective: Planning the Entire Curriculum

A high school teacher is supposed to teach specific curricula to specific classes each specific year. In order to achieve this goal and to teach properly all the curriculum contents, teachers must plan their teaching. The first step of this process is the breakdown of the entire curriculum to a list of contents, and the allocation of approximate number of teaching hours for each content.

Even though there are some traditions in the order of teaching a specific curriculum, some educators can learn from others' unique ideas or develop their own order of teaching that fit their experience and teaching believes. For example, the tradition is to teach loops before recursion, and in most cases arrays are also taught before recursion; nevertheless, there are different and opposite approaches as well (see, e.g., Bruce et al. 2005; Mirolo 2011).

In all cases, however, a yearly time allocation should take into the consideration different aspects of the learning-teaching process, both external-to-the-class factors and internal-to-the-class factors, combined with pedagogical factors.

The external-to-the-class considerations relate to the school organizational framework, and include factors such as, the weekly lesson schedule, the computer laboratory availability and the time allocated to lab experience (see Chap. 8), the number of tests that should be administered according to the school policy, and other school constraints (e.g., trips and special events).

The internal-to-the-class considerations relate to the characteristics of each specific class, and include factors such as, the number of pupils in the class and its general level, pupils' abilities and whether these abilities are homogenous or not, pupils' temperament, and more. Needless to say that the class characteristics are among the most important factors determining the teaching plan in general and, in particular, the teaching strategy a teacher chooses to apply.

While the role of class characteristics in the teaching plan is clear, we demonstrate the potential influence of external-to-the-class factors on the teaching plan. For example, different teaching plans should be set if out of the 3-weekly teaching hours, 1 h takes place in the computer lab or all lessons take place in the lab. The number of times that a teacher meets the class each week is also a meaningful factor. Teacher can meet the pupils for two lessons of 1 h in different days or for two successive hours; this different schedule influences the length of wasted time (entering the class, relating to homework, etc.) and the homework extent. In addition, educators should consider their pedagogical objectives and find the best way to integrate them into the teaching planning together with the different constraints.

The above external- and internal-to-the-class considerations influence the teaching of any topic, and in most cases do not change significantly over the school year. Some of them should be observed by the teacher during the first lessons in which she or he gets to know the class.

From this perspective, this time planning serves as the basis for the actual teaching process of the curriculum. Without such overall time planning, the teaching-learning process may be significantly affected and disrupted by local events in the class and/or in the school. For example, a teacher can teach the *if*-statement for too many hours from different reasons (e.g., several pupils face difficulties to understand it), and then, should teach quickly other important subjects, or even worse, may not be able to complete teaching the entire curriculum.

We note, though, that the yearly teaching planning should be flexible and subject to changes if needed. In practice, after teaching each specific topic in the class, a teacher should reflect on his or her original teaching plan and correct and update it accordingly. Clearly, this assertion is correct for teaching processes in other framework as well (e.g., the university).

11.2.2   Intermediate Level Perspective: Planning the Teaching of a Study Unit

A study unit is a sequence of lessons that aims to teach a (relatively) wide (computer science) topic. When a teacher plans the teaching of a study unit, he or she must relate to two issues that mutually influence each other: first, the content-knowledge and skills that learners are supposed to acquire, and second, the period of time, that is, number of lessons allocated for the teaching-learning process. The main target of this planning is to divide the teaching of a study unit into a sequence of lessons, including class activities and the assessment approach. Based on the realization of these two issues, a valuable planning process can be carried out.

A recommended process for a study unit planning includes the following actions:

1. List the concepts included in the said topic.
2. Review the experience of the computer science education community and its research literature about students' difficulties and misconceptions that may occur while learning the said topic (see Chap. 4).
3. Locate the concepts listed in Stage 1 along a time line, taking into the considerations the difficulties recognized in Stage 2.
4. Divide the list of concepts into a sequence of lessons, considering the total time allocated for the teaching of the said topic.

In the next steps, each lesson should be planned in detail, as is described in what follows.

11.2.3 Local Level Perspective: Planning a Lesson

A lesson plan can be described in different levels of depth. Novice computer science teachers usually need to plan their lessons in detail; after gaining some experience and building confidence in the fieldwork, lesson plans become less detailed.

A lesson plan should address the following aspects:

1. *The lesson's main objective:* What is the lesson about? What is its main message and its main content?
2. *The explicit content to be covered in the lesson:* Which subtopics are included within the lesson content? What are the pre-concepts pupils should learn and understand?
3. *The lesson operational targets:* What students should learn in the lesson? What skills should student develop? What concepts should students understand? What are the operational performances students are expected to gain?
4. *The lesson activities:* What kind of activities will be included in the class? (e.g., a trigger, explanations, set of exercises, a game, a group activity, peer assessment, an inquiry work in the computer lab, etc.)
5. *The learning assessment:* How will students' understanding will be monitored during the lesson?
6. *Pupils' homework:* What homework should a student perform to accomplish the lesson targets?

In addition, since the teacher knows his or her pupils' learning abilities, as well as the teaching route they passed together so far, the teacher should also consider the characteristics of the specific class that is going to learn the specific content. In other words, in addition to the difficulties and misconception mentioned in the literature with respect to the said topic, the teacher should be aware of the specific difficulties that his or her own learners face, and what may help them overcome these difficulties and improve their understanding. It implies that when a teacher considers how to vary the teaching methods and class activities employed during the year, he or she should be aware to what works and what does not work for each specific class.

11.2.4 Building Concept Understanding in a Spiral Gradient Manner

In some cases, it is better to build the understanding of a concept in spiral gradient manner. There are places where we apply this principle quite naturally, for example, in the case of the concept of modularity. Specifically, when exploring the notion of modularity, educators start from simple tasks, and gradually progress to complicated ones. However, for some meaningful concepts this approach is not applied naturally. In such cases, it is recommended to carefully plan the teaching process, exposing the learners to the essence of the concept gradually along the curriculum. This kind of interpretation requires a broad view of the whole curriculum planning, integration of exercises that enable to overwhelm the concept along the curriculum, and conduct discussions of the concept. This methodology enables students to gradually build their understanding in the spirit of the constructivist learning approach.

A good example for such an approach is with respect to the concept of algorithm efficiency, which is usually referred to as "hard to accomplish". This is exactly the reason why efficiency should be presented gradually, instead of presenting it for the first time much later by sort algorithms, as is done in most cases. In the case of algorithms efficiency, the spiral gradient teaching could be done, for example, by the following exercises:

1. When teaching conditions, one of the very first exercises is "Find the biggest number out of three numbers." Even in this simple case, we can discuss the number of conditions used in different solutions and show a solution that uses only two simple conditions.
2. When teaching loops, one of the common exercises is: "Print all the divisors of a number n." Here we have the opportunity to look at a loop that repeats till n—looking at all the numbers smaller than n; a loop that goes till $n/2$—looking at the optional dividers; and a look at loop that goes till $\sqrt{n}$ —till the largest possible divider.
3. When teaching arrays, we usually use the next exercise: "For n input numbers print the most frequent number." Here, we must use double loops, while the inner loop can be twice smaller on average; when proceeding to counter arrays we can change the exercise and talk about a case when the range of the values is between 1–100 and use only one loop.

Such exercises, as well as the discussion that follow the presentation of their solutions, should be integrated along the curriculum.

More exploration on the teaching of efficiency to novices can be found at Ginat (1996).

11.3 Illustration: Teaching One-Dimensional Array

We now demonstrate the teaching planning of one study unit by focusing on the topic of one-dimensional arrays (Sect. 11.3.1), and the teaching planning of one specific lesson—the first lesson about arrays (Sect. 11.3.2). Needless to say, other possible planning processes exist and the presented planning is just one option among many.

11.3.1 Planning the Teaching of a Study Unit about One-Dimensional Array

The teaching planning of the study unit about one-dimensional array is carried out in what follows according to the stages presented in Sect. 11.2.2.

1. *List the concepts included in the said topic:* This list contains different aspects and different scopes of one-dimensional array. The following list, formulated informally on purpose, reflects this approach:
   - The need for the array structure.
   - When is it necessary to use arrays and when it is not?
   - The structure of a basic array: an ordered collection of cells from the same type with one shared name and a unique index for each element.
   - The distinction between an array-cell index and an array-cell content.
   - Understanding that each array-cell functions as any other variable.
   - Array boundaries.
   - Basic array scans (for insert, retrieve values, etc.).
   - The array representation in a specific programming language: array declaration, syntax, memory structure (e.g., in Java, an array is an object and the array length is an attribute of the object).
   - Array patterns (patterns used for "classic" algorithms such as find the max value, print all items/elements, etc.).
   - Different tasks on arrays that implement different algorithmic approaches with different logic complexity (e.g., find the max value, average value, number of elements that are larger than the average, frequency of a given value, the most frequent value, changing values according to different rules, find sub-sequences, and more).
   - Subtasks using arrays (parameter passage and arrays as a returned value of a function).
   - Tasks that involve array building (building a sub-array of a given array according to a specified condition, array union, array intersection, array subtraction).
   - Array of counters.
   - Array of accumulators.
   - Array of objects (in the relevant programming languages).
   - Search in arrays (linear or binary).
   - Sorting arrays (insertion sort, selection sort, bubble sort, recursive sort like quick-sort).
   - Merging of arrays (e.g., sort-merge).
   - The efficiency of different array algorithms.
2. *Review the experience of the computer science education community and its research literature about students' difficulties and misconceptions that may occur while learning of the said topic:* We present two lists related to learners' difficulties and misconceptions while learning arrays: the first one presents experiences of practitioners in computer science education using a casual ter-

minology; the second list presents a list which is based on the computer science education research literature.

Experiences of computer science educators:
- Confusion between the index and the content of an array-cell.
- *Misconception:* When an array is used, all its cells should be scanned.
- *Misconception:* Including the array variable in an output statement displays all the array cell contents.
- When an array of objects is used, difficulties in understanding the need to construct each object (even after the array of objects itself has already been constructed).
- Difficulty to understand the algorithmic role of an array cell (e.g., a regular value, a counter, an accumulator).
- *A frequent mistake:* exceeding the array index beyond the array size.
- *A frequent mistake:* loss of a value of an array-cell due to overwriting (e.g., in the process of array sorting).
- *A frequent mistake:* When building a new array, leave empty cells in the "middle" instead of writing values successively (e.g., in intersection).

Examples from the computer science education literature:

Table 11.1 presents seven examples, taken from the computer science education literature, that discuss students' difficulties and misconceptions with respect to learning and teaching the concept of array. For each reference, we suggest a possible implementation for the teaching planning of the study unit of arrays.

3. *Locate the concepts listed in Stage 1 along a time line, taking into the considerations the difficulties recognized in Stage 2.*
4. *Divide the list of concepts into a sequence of lessons, considering the total time allocated for the teaching of the said topic.*

These stages (the actual allocation of each topic on a time line and its division into successive lessons) should be carried out according to learners' age and abilities, the total time allocated for the study unit, the number of lab lessons, etc. In addition, it is important to include evaluation points along this time line and to decide what kind of evaluation is appropriate for this case (e.g., tests, formative evaluation, portfolio; see Chap. 10).

11.3.2   Planning the Teaching of the First Lesson about One-Dimensional Array

The process of lesson planning is illustrated by the first lesson in the study unit of one-dimensional array, and is carried out according to the consideration of the aspects presented in Sect. 11.2.3. We assume that (a) the learners are high school pupils, (b) the time allocated for this lesson is one and a half hour, and (c) the lesson takes place in the computer lab.

Table 11.1 Examples from the computer science education literature and their possible implementations for the teaching planning of a study unit of arrays

Examples of computer science education research works that discuss learning and teaching array	Possible implementation for teaching planning
A game that teaches loops and arrays in an interactive and visual way (Eagle and Barnes 2008)	Use the suggested game, or, alternatively, apply the game rational and ideas for the development of another game
Array algorithms are defined as functional algorithms where each step of the algorithm results in a function being applied on an array, producing an array as a result. The rationale for teaching array algorithms is given together with an example, which shows that array algorithms sometimes lead to surprising results (Howland 2005)	Learn this kind of algorithms, and see what can be adapted from the rational of its teaching and from the example
Examination whether the shift to the object-oriented programming and the application of the objects-first or objects-early approach to CS1 requires to reevaluate the following pedagogical question: What should the first data structure students are exposed to be? Is it an array or another kind of collection, for example, a map? (Ventura et al. 2004)	A critical thinking is encouraged with respect to the first data structure to teach
Examination whether the shift to object-oriented programming techniques calls for a significant shift in the approach of teaching recursion. Should simple recursive structures, such as linked lists and methods that process them, be introduced before procedural examples? (Bruce et al. 2005)	A critical thinking is fostered with respect to data structures from the recursion point of view (see also Chap. 12): What data structures should be taught earlier? Can the chosen data structure be manipulated by recursion procedures?
Examination which algorithm of array sorting is better to introduce first, and what difficulties learners may face (Nieminen 2006, 2008)	Thinking about the order of teaching different array sorting algorithms and being aware of possible obstacles
A blog resource about the history of sorting algorithms (Abhiram 2009)	This resource can form a basis for a web-based activity (see Sect. 8.5 Using the Internet in the Teaching of Computer Science). This is an example (among many others that can be found on the web) of using an online resource
Beginning students performed significantly better when using of test-driven development (TDD) approach using WebIDE. Data presented specifically on arrays (Hilton and Janzen 2012)	This resource can form a basis for an inquiry lab-activity (see Sect. 8.3 Lab-First Approach)

1. The lesson's main objective is to let the pupils realize the basic need for arrays and to facilitate basic manipulations of arrays.
2. The subtopics to be addressed in this lesson are the basic structure of array, a cell index, cell content, and basic scans of array.
3. The lesson operational targets are:
   - Learners become aware of simple kinds of tasks for which arrays are needed.
   - Learners know how to access an array cell.

- Learners understand that each array cell functions like a variable.
- Learners distinguish between the expressions *a1*—a single variable, *a*[1]—the value of the array cell whose index is 1, and *a*[*i*]—the value of an array cell whose index is *i*—the value of a variable *i*.
- Learners distinguish between the expression *i* as an index and *a*[*i*] as a cell content.
- Learners know how to scan an array.

4. The following activities to be facilitated in the lesson were designed according to the following guidelines:
   - The lesson addresses arrays from two perspectives: conceptual and practical. It is important to illuminate these two perspectives to the pupils.
   - The lesson involves different kinds of activities: a trigger, work in pairs, a class discussion, and a lab activity.
   - The lesson plan takes into the consideration the expected difficulties mentioned in Sect. 11.3.1 and attempts to reduce the level of abstraction (when needed) to help learners overcome these difficulties.

The lesson activities are presented in what follows along the lesson time line together with their main idea. In some cases, we also specify the assignment.

- *Presentation of a trigger to the class plenum:* The learners are asked to solve a problem whose solution requests an array and then, in pairs, to check the pair's solution. Since at this stage, they are not familiar with the concept of array, they cannot solve the problem and therefore, it is expected that they will feel some inconvenience.

  An example of such a trigger is: Write a program that prints the number of pupils whose grade is bigger than the class average for a class of 34 high school pupils.

  One solution that pupils may suggest is to read the grades twice: first, in order to calculate the average, and second, in order to count the number of grades which are bigger than the average. In this case, a common computer science convention should be added to the question formulation: In computerized systems, it is common to enter data only once. The learners will be given additional time to deal with the new constraint, and to realize that it is not possible to solve the problem with this restriction.

  Another solution that can be suggested by the pupils is to define 34 different variables, one for each grade, namely, *a1, a2,..., a34*. In this case, the learners should be directed to observe what this solution includes 34 input statements, 34 addition statements and 34 condition statements. In this case, additional common computer science convention can be added: Computerized systems should be able to generalize the problem; in this case, what will we do if there are 100 learners? After all, we wish the computer to work for us and not us to work for the computer. As in the previous case, the learners will get extra time to deal with the new constraint, and again, to realize that it is not possible to solve the problem with this restriction.

- *Class discussion:* A discussion is facilitated about the need to have a data structure that enables to process any amount of data, but at the same time enables simple manipulation.
- *Work in pairs:* Pairs of pupils solve together a worksheet. The worksheet contains about ten different problems. For each problem, the learners should determine whether its solution requires an array or not. The discussion about the worksheet is summarized in the class plenum.
- *Presenting the basic structure of an array:* array name, array-cell index, the indexes range, array-cell content (the use of $a[3]$ versus the use of $a[i]$ or $a[i*2]$).
- *Practicing in the computer lab:* Pupils are asked to work on a worksheet that guides them watching how arrays can be used within one of the animation environments (see Sect. 8.4). This activity aims at enhancing the learners' understanding of the new structure, and how it can be used and scanned. This aim is achieved by presenting the data structure memory organization in these environments in such a way that reduces the level of abstraction.

5. Learners' understanding can be assessed during the lesson in several opportunities, for example, the discussions about the trigger and about the worksheet, the work on the lab activity, and the questions included in the worksheet, and, finally, homework.
6. The pupils' homework is a worksheet with different types of questions (see Chap. 9). The questions should address the different ideas taught in the lesson, both conceptually (e.g., problem analysis) and practically (e.g., code execution in one of the animation environment).

11.3.3 Illustration Summary

This demonstration of teaching planning reflects the complex and multifaceted mission of teachers in general and of computer science teachers in particular. The detailed teaching planning illustrates also how different topics addressed in the different chapters of this guide are merged in practice: problem solving (Chap. 5), learners' difficulties and misconceptions (Chap. 6), computer science education research (Chap. 4), different class activities and class organizations (Chaps. 2 and 7), lab activities (Chap. 8), and types of questions (Chap. 9).

Each student in the MTCS course, when becoming a computer science teacher, will teach topics that are included in his or her state/country specific curriculum. Several of the main targets of the MTCS course (e.g., students' exposure to a variety of teaching tools and increasing their familiarity with a set of considerations that will guide them in the planning process of their teaching and in the actual teaching process) are aiming at preparing them for this task. These targets are achieved, for example, by asking the students to develop activities for learners by using a variety of teaching methods (see Chap. 7). This kind of work, enables students to acquire and improve their understanding both with respect to the teaching methods and the taught concept itself.

In this spirit, the Activities 87–90, to be facilitated in the MTCS course, aim to explore the issues addressed in this chapter. In addition, the different considerations presented in the chapter should be integrated in other course activities when appropriate, especially the planning teaching of a lesson, which should be practiced along the course several times.

Activity 87: Dividing a Computer Science Topic into Components

In this activity, the students work on one specific computer science topic, for example, the *if* statement, loops, or lists. The work is carried out first individually, then in groups, and finally, in the course plenum.
    Specifically, the students are asked to:
1. Choose a topic from the curriculum they intend to teach.
2. List the subtopics of the said topic.
3. Order the subtopics on a teaching time line and to explain what, in their opinion, should be taught first, what subtopic/s should be taught after that, the teaching of what topic can be postponed, etc.
4. Specify difficulties learners are expected to face. If needed, the students should change the ordered list of subtopics presented in stage (2) in a way that helps learners overcome these difficulties.

Activity 88: Time Allocation, Team Work

This activity can be facilitated either following Activity 87 or independently. A list of subtopics of some topic that should be taught to a specific class is presented. The students are asked to determine the time needed to teach each subtopic. The term teaching should include the actual teaching of each subtopic, learners' practicing of each subtopic, and the assessment of learners' understanding.
    After the students work on the activity in groups, the different suggestions of each group are presented together with the group's considerations. It is important to highlight differences between the groups and the reasons that led them to determine different time allocations.

Activity 89: Plan a First Lesson about a Topic/Subtopic

The students are asked to plan the first lesson of a topic which is new to a specific class of learners. It is optional to facilitate this activity with respect to a specific teaching method discussed in the course. For example, when different usages of the Internet in computer science education are addressed in the MTCS course (see Chap. 8), students can be asked to plan the first lesson about array sorting based on learners' exploration of Internet simulations.

Activity 90: A Comprehensive Teaching Planning of a Study Unit

The objective of this activity is to let each student in the MTCS course delve into the details of the actual construction of a teaching plan for a full study unit. Specifically, each student chooses a topic from the high school curriculum, analyzes it, and plans a study unit for it. As can be seen, in addition to practicing teaching planning, it enables the students in the MTCS course to express knowledge related to different topics studied in the course as well as in other courses. Since it is a comprehensive activity, it can be served as a full-semester/term summary work that should be carried out during a long period of time (Ragonis and Hazzan 2008).

It is recommended that the course instructor leads the students to choose different computer science topics for this work and to publish these works in the course website so that all the students could benefit from their colleagues' work. It is also important that the course instructor guide the students during their work since they may face dilemmas and take decisions. The instructor's feedback, based on his or her own experience and knowledge, can support the students in this work preparation process as well as in their general professional development.

The final work that the students submit should include the following sections:

1. Concepts and contents to be taught.
2. Difficulties expected to be encountered by learners when learning the selected topic. This part should be based on at least one research paper that addresses the topic.
3. A high-level division of the topic into lessons, specifying the recommended teaching sequence.
4. Designing a full lesson plan for two consecutive lessons, specifying the pedagogical principles that should guide the teaching process. The two detailed lesson plans should include the following components:
   - Full lesson plan, presenting the lesson's objectives and the lesson opening, development, and ending
   - Description of the activities/questions/tasks/exercises, etc. included in the lesson
   - Description of the teaching methods to be used in the lesson, e.g., frontal teaching, working in small groups, individual learning, investigative and discovery activity, games, class discussions, etc.
   - Description of the teaching aids to be used in the lesson, for example, overhead projector, posters, models, computer, simulations, animations, and the computerized learning environment
   - Suggestions for learners' evaluation after the two lessons, solutions of the evaluation tasks, and an evaluating rubric for the task evaluation (see Chap. 10)

5. Written reflection on the entire development process.

It is recommended to summarize this activity in one or two lessons of the MTCS course, in which each student presents his or her work for 20–30 min. The presentations can include two parts: (1) a summary of the individual work and (2) introduction of a short excerpt from the planned lessons, which can be taught by each student to his or her peers, together with a description of the considerations that guided the development of this segment of the lesson. It is also suggested that the course instructor encourages the students to reflect on the process they went throughout the fulfillment of this activity.

References

Abhiram (2009) History of sorting algorithms. Blog spot. http://abhiramn.blogspot.com/2009/07/history-of-sorting-algorithms.html. Accessed 10 May 2010

Bruce KB, Danyluk A, Murtagh T (2005) Why structural recursion should be taught before arrays in CS 1. ACM SIGCSE Bull 37(1):246–250

Eagle M, Barnes T (2008) Wu's castle: teaching arrays and loops in a game. ACM SIGCSE Bull 40(3):245–249

Ginat D (1996) Efficiency of algorithms for programming beginners. SIGCSE Bull 28(1):256–260

Hilton M, Janzen DS (2012). On teaching arrays with test-driven learning in WebIDE. In: Proceedings of the 17th ACM annual conference on Innovation and technology in computer science education (ITiCSE '12). ACM, New York, NY, USA, 93–98

Howland JE (2005) Array algorithms. J Comput Small Coll 20(4):229–235

Mirolo C (2011) Is iteration really easier to master than recursion: an investigation in a functional-first CS1 context. In: Proceedings of the 16th annual joint conference on Innovation and technology in computer science education (ITiCSE '11). ACM, New York, NY, USA, 362–362

Nieminen J (2006) Bubble sort as the first sorting algorithm. http://warp.povusers.org/grrr/bubblesort_eng.html. Accessed 15 May 2010

Nieminen J (2008) Bubble sort misconceptions. http://warp.povusers.org/grrr/bubblesort_misconceptions.Html. Accessed 15 May 2010

Ragonis N, Hazzan O (2008) Disciplinary-pedagogical teacher preparation for pre-service computer science teachers: rationale and implementation. In: Mittermeir RT, Syslo MM (eds) Information education—supporting computational thinking, lecture notes computer science 5090, ISSEP 2008. Springer, Berlin, pp 253–264

Ventura P, Egert C, Decker A (2004) Ancestor worship in CS1: on the primacy of arrays. 19th Ann. ACM SIGPLAN OOPSLA Conf.: 8–72

Integrated View at the MTCS Course Organization: The Case of Recursion

12

Abstract

This chapter presents an optional organization theme for the Methods of Teaching Computer Science (MTCS) course around the concept of recursion. Based on the active learning-based teaching model, a series of themes is suggested, each one highlights a different pedagogical perspective. The themes are: classification of recursive phenomena (a nonprogramming task), the "leap of faith" approach, models of the recursive process, research on learning/teaching recursion, how does recursion sound? (the case of trees and fractals), evaluation (a nonprogramming project and a test construction), and a list of additional activities that illustrates that recursion can, indeed, be the focus of almost any topic discussed in the MTCS course. Each theme is accompanied with activities devoted to recursion to be facilitated in the MTCS course.

12.1 Introduction

In the introduction to this guide, we mentioned that the Methods of Teaching Computer Science (MTCS) course should not necessarily follow the order of the chapters as they are presented in this guide. In this spirit, this chapter offers an alternative approach according to which the organization of the MTCS course is based around one (or more) central computer science concept(s). By doing so, this chapter illustrates an optional organization theme for the topics presented in Chaps. 2–11 of this guide. Section 14.2 suggests two additional possible syllabi for an MTCS course.

Specifically, this chapter reviews the guide's chapters through the lens of recursion[1]—one of the central computer science concepts. Though the ideas presented in

---

[1] Recursion is only one candidate for such course organization. Other central computer science themes, such as abstract data types or CSE research, may also be used for the same purpose.

© Springer-Verlag London Limited 2014
O. Hazzan et al., *Guide to Teaching Computer Science*,
DOI 10.1007/978-1-4471-6630-6_12

this chapter can be used by all computer science educators, our focus is placed on the MTCS course.

Recursion is chosen for this purpose for four main reasons. First, recursion is one of the central computer science concepts included in almost all introductory computer science courses. Second, recursion is linked to many other computer science topics and fields, and therefore, discussions about this concept that take place in the MTCS course may provide opportunities to highlight other computer science topics from a different, and sometimes less familiar, angle. Third, recursion has interdisciplinary relations to other areas in our life (including science and art); this fact highlights its advantages from a pedagogical perspective, and from a cognitive perspective, and plays a significant role with respect to the learning of recursive processes (see, e.g., Hofstadter 1979). Last, but not least, recursion is an interesting and exciting concept to learn and explore.

Recursion is a central computer science idea mainly because it enables to describe complex algorithms and data structures in a simple and elegant manner by applying the idea of self-reference to programming (Harvey and Wright 1999). It is usually defined as a programming tool or a programming techniques; recursion, however, is also relevant for the examination of objects structure.

The idea of recursion is also applied for definitions. In a recursive definition, the defined concept is part of the definition itself (Gersting 1996). Specifically, a typical recursive definition has two parts: One part describes the simplest base case (or cases); a second part describes how to reduce complex cases into a simpler, yet similar, case(s), by a set of rules which reduce all cases to the base case(s). Accordingly, one part of a recursive algorithm describes actions to be executed in the simplest case, and another part describes the recursive call (in which, the algorithm activates itself). In the case of data structures, one part of the recursive definition describes the simplest structure and another part describes how the entire structure includes a simpler version of itself.

This chapter is based on the Active-Learning-Based Teaching Model, introduced in Chap. 2. In this chapter, we suggest a series of themes, together with associated activities all devoted to recursion when each theme highlights a different pedagogical perspective, as is described in what follows:

- *Classification* (Sect. 12.2): This theme focuses on recursion as a soft idea (see Sect. 3.7), suggests a teaching method which is based on classification tasks (see Sect. 7.2.5) and highlights the notion of diversity (see Sect. 3.5.2).
- *Leap of faith* (Sect. 12.3): This theme introduces a teaching method that can help learners develop their formulation skills of recursive descriptions (without necessarily understanding how the recursive descriptions actually work). The activities presented with respect to this theme relate to learners' difficulties (see Chap. 6) and teaching planning-related issues (see Chap. 11).
- *Models of the recursive process* (Sect. 12.4): In this section, the focus is placed on models that can support learners' understanding of the recursive process. The activities presented in this section relate to teaching methods (see Chap. 7) and to learners' conceptions (see Chap. 6).

- *Research on learning/teaching recursion* (Sect. 12.5): This theme examines research findings related to the learning and teaching of recursion. The activities presented in this section relate to research in computer science education (Chap. 4) and learners' conceptions (Chap. 6).
- *How does recursion sound?* (Sect. 12.6): This theme focuses on lab-based teaching (see Chap. 8) with respect to recursion, and suggests a music-based activity for enhancing learners' understanding of trees and fractals.
- *Assessment* (Sect. 12.7): The focus of this theme is on evaluation aspects (see Chap. 10) with respect to learning recursion. Two activities are suggested: the first activity examines a nonprogramming project as an evaluation tool for learners' understanding of recursion[2]; the second activity deals with test construction.
- *Additional activities* (Sect. 12.8): The last theme includes additional activities to be facilitated in the MTCS course, all of them relate to recursion learning and teaching.

12.2   Classification of Everyday Objects and Phenomena: The Case of Recursion

In Sect. 7.2.5, classification is introduced as a teaching method in computer science education. Activity 91 illustrates this teaching method with respect to recursion. The first two stages aim at strengthening the students' own understanding of the concept of recursion; the last two stages discuss this kind of activities from a pedagogical perspective.

Activity 91: Classification Activity in the Context of Recursion

- Stage A: Classification activity, work in small teams
- The activity is based on the analysis of recursive phenomena taken from various fields such as art, music, literature, and mathematics (see Levy and Lapidot 2000). As mentioned in Sect. 3.5.2, triggers encourage the expression of *diverse* perspectives and ways of thinking. Yet, this particular trigger of the classification task was found to be most successful with its mental construction processes, rich discussions, and powerful analogies for further learning (Levy and Lapidot 2000).
- The students are not informed about the nature of these phenomena and their analysis should be based on their daily life experience. Specifically, the students are asked to work in small teams and to classify these instances according to their own criteria. Clearly, there is no correct classification; these criteria, however, are often found to be important constructs in learners'

---

[2] This activity relates also to Chap. 3 (what is CS and its relations to other fields, diversity of learners).

mental construction process of the concept of recursion[3] (Levy and Lapidot 2000).

- The students are also asked (a) to expand their classified sets by adding new instances to each set, (b) to give a title to each set, and (c) suggest a title for the whole page. The specific instructions are presented in Table 12.1.

Table 12.1 Classification task

Worksheet: Classification task
Levy and Lapidot (2000) present a page with 15 phenomena (see appendix A of the article). Choose your own criteria and categorize/classify these images into sets. An image can belong to several sets. For example, image X can be included in set A because it satisfies the 'a criterion' and to be included in set B because it also satisfies the "b criterion."
Add a new instance (not from the given page) to each set.
Give a title to each set.
Give a title to the whole page.

- Stage B: Class discussion
- After the teams worked on their classification, each team shares its categorization with the rest of the class. This sharing process can be performed in different forms. Here are several options: (1) a group presents the instances of a specific set and the whole class should guess their classification criterion, (2) a group presents its additional new instance to one of the sets and the class should guess which of the other instances presented in the worksheet belong to this set, (3) a group presents their title for one of the sets and the whole class should guess which instances belong to that set, and (4) a group presents 3–5 instances that cannot belong to one of its sets and the class should guess what the criterion was according to which items were included in that set (i.e., what was the classification criterion).
- During the discussion, the instructor should encourage a reflective discourse, offer generalizations, and present the formal terminology with respect to recursion related to the mentioned constructs. Learners, in general, and the prospective computer science teachers in the MTCS course, in particular, are often exposed in this discussion to new concepts and to different ideas and perspectives offered by other groups. This exposure, in turn, encourages their reconsideration of their previous perspective at recursion.
- The instructor summarizes the main concepts related to recursion that have been introduced in this discussion, and encapsulates them under the umbrella of one concept—recursion. As a preparation for the next stages of

---

[3] This observation is not surprising since the phenomena were chosen very carefully so that they represent recursive structures and entities.

this activity, in which classification tasks are examined from a pedagogical perspective, the students are told that this page aims at introducing the concept of recursion to computer science learners who are *not* familiar with this concept.

- Stage C: Build a classification page, homework, and class discussion
- Before homework is presented, it is recommended to discuss with the students what other computer science concepts can be introduced to computer science learners by classification tasks and what characterizes these concepts.
- As their homework assignment, the students are asked to build another classification page for a different computer science concept. As mentioned in Chap. 7 (Sect. 7.2.5), good candidates for such concepts are data structures, control structures, and abstraction.
- If time permits, it is recommended to dedicate an additional lesson in which the students present their classification pages. In their presentation, they should be encouraged to address how they chose the concept for which the page was constructed and how they selected the instances included in their page. These presentations should be followed by a class discussion which emphasizes the teacher's perspective. This discussion can focus on advantages and disadvantages of classification tasks as a pedagogical method for introducing new concepts to computer science learners. The instructor can explicitly direct the students to reflect on what they learned during their work on the classification page about recursion. It is worthwhile addressing in this discussion concepts such as classification, generalization, abstraction, mental construction, constructivism, active learning, group work, alternative ways to introduce recursion (or other computer science concepts), and visualization (see also Chap. 7).
- Stage D: Summary: read a paper, homework
- This topic can be summarized by the following homework:
- Read the paper Levy 2001.
- Choose at least one aspect of the classification activity that was not discussed in the class and explain its importance from a pedagogical perspective.
- Among the research findings presented in Sect. 5 of the paper, choose one finding that surprised you and explain why it surprised you.

12.3   Leap of Faith

Once learners are familiar with the concept of recursion, for example, by working on the classification task, programming aspects of recursion can be addressed. This transfer, however, should be done very carefully since there is an agreement that understanding recursion sets cognitive challenges for novice computer science learners (George 2000).

At least three factors contribute to these difficulties: (1) the gap between the recursive (simple) algorithm and the recursive (complex) execution process: In order to understand recursion, one must distinguish between the program (or method) listing and its recursive process (algorithm execution) and further, these two instances require different kinds of understanding, based on different cognitive models and abilities, (2) learners' faulty mental models of the recursive execution process, and (3) pedagogy of teaching recursion. Learners' difficulties of learning recursion are elaborated in Sect. 12.5.

According to Leron (1988), in the first stages of teaching recursion, it is preferable to concentrate on the relations between the algorithm and its product or output (the result of the algorithm execution), rather than on the relations between the algorithm and the process it invokes. This pedagogical suggestion is based on the recognition that learners should understand first how recursive phenomena can be described recursively and only then, to cope with the complex recursive execution process. Harvey (1997) and Harvey and Wright (1999) call this teaching approach *a leap of faith*.

In general, a leap of faith refers to one's belief in the existence of a phenomenon that cannot be touched or proved and for which no evidence exists. In our case, *the leap of faith* method is associated with the assumption that an algorithm one writes works properly. It implies that *the leap of faith* method guides pupils to write recursive descriptions even if they do not fully understand (yet) why and how this "magic" works. It is recommended to start implementing this approach with respect to recursive shapes (e.g., fractals or trees) or other recursive phenomena; then, once pupils are able to write recursive descriptions for these instances, they can use this approach for writing recursive functions as well (again, prior to their fully understanding of the function execution process).

Teaching *the leap of faith* method to learners is not a trivial matter and requires some practice. Activity 92 aims at preparing the prospective computer science teachers mastering this approach (Lapidot et al. 2000).

Activity 92: Mastering the Leap of Faith Approach
- Stage A: Toward the leap of faith approach, team work
- The students in the MTCS course are given the worksheet presented in Table 12.2 to work on.

Table 12.2 Leap of faith worksheet

Worksheet: toward the leap of faith approach
In this worksheet, you are asked to work on four short tasks and then to design similar tasks.
Worksheet, part A: Tasks

Table 12.2 (continued)

Worksheet: toward the leap of faith approach

First task: A recursive description of a visual figure (binary tree) is given. Draw the described figure for level 3.
The basic binary tree is of level 1, and it looks like a V-shape.
A binary tree of level N is a V-shape with a smaller binary tree at each edge of the V-shape.

Second task: A recursive description of a valid expression is given. Write as many valid expressions as you can.
The basic valid expression is the letter C.
A valid expression begins with the letter A, proceeds with a shorter valid expression, and ends with the letter B.

Third task: Represent the following sand clocks by a recursive description, similar to the two descriptions presented above.

$$
\begin{array}{cc}
55555 & \\
4444 & \\
333 & 4444 \\
22 & 333 \\
1 & 22 \\
22 & 1 \\
333 & 22 \\
4444 & 333 \\
55555 & 4444
\end{array}
$$

Fourth task: A series of numbers is given. Write two formulas for the nth element of the series: one formula should be recursive; the other formula—not recursive.

$$2, 5, 8, 11, 14, 17, 20....$$
$$A_1 =$$
$$A_n =$$

Worksheet, part B: Design similar tasks

What is the structure of a recursive description?

Design at least one additional task for each kind of the four tasks presented in part A.
The first task: translating a recursive description into a visual figure
The second task: translating a recursive description into a textual (nonvisual) expression
The third task: translating a visual figure into a recursive description
The fourth task: representing a series of numbers by a recursive formula and by a nonrecursive formula

Design a new kind of tasks that deal with recursive descriptions and are different from those presented in part A

- Stage B: Class discussion
- Once the students finish working on the two parts of the worksheet, it is recommended to facilitate a whole class discussion, which clarifies the purposes of the tasks presented in part A, that is, getting familiarity with

the structure of a *recursive description* and practice the creation of recursive descriptions. It is also important to emphasize the structure of a recursive description (that is, naming the recursive phenomena, a base case, and a recursive call). At this point, the meaning of the leap of faith approach can be explained.

- If time permits, it is also recommended to start discussing with the students the planning of teaching recursion. Though there are several relevant topics with which the students are not familiar with yet (e.g., research on recursion learning), a preliminary discussion, which emphasizes the important role of nonprogramming activities in learning recursion (as the ones presented above) can start at this point. The role of such activities should be associated with the complexity involved in learning the concept of recursion.
- Stage C: Summary (read a paper, write a program), homework
- The students are given the following tasks:
  1. Read one of the following papers. Then, choose at least one aspect of learning and teaching recursion that the paper addresses and that was not discussed in the class and explain its importance:
     - Leron U (1988) What makes recursion hard? Proceedings of the 6th International Congress on Mathematical Education (ICME6), Budapest
     - Levy D, Lapidot T (2000)Recursively speaking: analyzing students' discourse of recursive phenomena. Proceedings of the 31st SIGCSE technical symposium on computer science education, Austin, Texas, pp 315–319
  2. Write a Java program for the Fibonacci series by following the "leap of faith" method. Explain the stages you followed (that is, how you used the "leap of faith" approach while building your program).

12.4   Models of the Recursive Process

In this section, we focus on models of the recursive process. We assume that at this stage learners got some experience with respect to the formulation of recursive descriptions (Sect. 12.3) and therefore, they can proceed to the learning of the recursive *process* (i.e., the execution of an algorithm).

Due to the complexity gap between the relatively simple recursive algorithm[4] and the complex recursive execution process, the recursive process is considered to be one of the most difficult issues for understanding with respect to recursion.

All learners build mental models and use them as part of their learning processes. In the case of recursion, research reveals that many learners hold a faulty model of

---

[4] This simplicity refers to the fact that a recursive algorithm is usually short with only few instructions. This fact stands in contrast to the complex process invoked by a recursive algorithm.

the recursion process (see, e.g., Götschi et al. 2003; Kahney 1989; Wu et al. 1998). Sometimes, even viable models, such as the copies mental model, do not help learners to execute recursive algorithms correctly (Scholtz and Sanders 2010).

Therefore, suitable tracing models are needed to help learners capture the essence of the recursive process. Two models are presented here: the Little People model and the Top-Down Frames model. Both models are demonstrated by the following mystery method.

```
public static void mystery (int n) {
    if (n < 1)
                System.out.println ("finished");
    else {
                System.out.println (n);
                mystery (n-1);
                System.out.println (n);
    }
}
```

This example is chosen for several reasons. First, in order to ease the tracing of the process, all the instructions included in this method are based on printings. Second, the example is a procedure (and not a function), because it is more difficult to explain the return process included in a recursive function (like factorial). Third, one of the difficulties learners face with respect to understanding recursion is the computer's behavior when a recursive execution ends; therefore, it is recommended to start with a full recursion (not a tail recursion) since its flow is easier to understand.

12.4.1   The Little People Model

The little people model is a powerful model that helps learners actually see the execution of a recursive process. This model not only provides a tool that simulates the recursive process but is also based on learners' active participation in the creation of the recursive process, and at each step enables them to predict the next step(s). In his *Computer Science Logo Style* book, Harvey (1997) explains the model in detail and illustrates it with detailed examples. In what follows, it is described briefly.

The model assumes that a large community of little people[5] exists inside the computer, where (a) each person is an expert in the execution of a specific method/program, and (b) more than one little person can have the same expertise. Thus, for our example, there are *mystery* specialists. When a method is invoked, one little person, who is an expert in its execution, is called by another little person to execute it. The little person wears a vest with pockets; the number of pockets equals the number of the inputs of the program/method that the little person is an expert of its execution. In our case, any *mystery* expert has one pocket. Each pocket can contain

---

[5] Harvey calls these little people elfs or specialized doctors.

a value of one variable. In addition, each pocket has a tag name that presents the name of the parameter, attached to its inner side.[6] We also assume that all the little people are familiar with the basic Java instructions.[7]

Activity 93, to be facilitated in the MTCS course, illustrates the Little People Model and further, enables to address it from a pedagogical perspective.

Activity 93: Pedagogical Examination of the Little People Model
- Stage A: Demonstration of the Little People model
- The stage is based on an active role play of executing *mystery* (3) by the students. Each step is presented in a new bullet:
  - The chief person (the MTCS course instructor or a student who is the manager) calls one student—Anna, for example, one of the *mystery* experts, to perform *mystery*(3), puts *3* in her pocket (*n*) and gives her a chalk.[8]
  - Now, Anna should check the *if* statement: She looks at the value of *n* in her pocket and checks whether $3 < 1$. Since the condition is not satisfied, Anna continues to the *else*-clause.
  - Since Anna knows the basic instructions of Java, she knows how to perform System.out.println. She looks again in her pocket, sees that the value of *n* is *3* and prints *3* (on the blackboard, for example).
  - At this stage, Anna should perform *mystery(n−1)*; since this is not a basic Java instruction, she needs help from another little person and she calls Burt.
  - In the class, at this point, the students are encouraged to suggest Anna what to do and a short discussion can take place about the fact that Anna cannot perform *mystery(2)* since she is still busy with her execution of *mystery(3)*.
  - Anna hands the chalk to Burt, looks at her pocket for *n*'s value, calculate *n−1*, puts 2 in Burt's pocket, and goes standing at the side of the classroom waiting for Burt to finish his job.
  - Burt should check if $n < 1$. So he looks for *n*'s value in his pocket (*n*) and then checks if $2 < 1$. Since the condition is not satisfied, Burt turns to the else-clause. He knows how to perform System.out.println, so he prints 2 on the blackboard. Then, in a similar manner to Anna's, he should perform mystery(*n−1*) and calls Carl to perform this job. Burt

---

[6] The name tags are attached to the inner side of the pocket to emphasize the fact that the names of an expert's variables are not exposed to other little people (indeed, they should not know these names); they should know only the number of pockets and the kind of thing (types of variables) that can be put inside them.

[7] In the full implementation of the little people model there are experts also for the System.out. println command and the if statement; this assumption is not necessary for our discussion.

[8] The chalk holder represents the active actor at each stage of the role play.

gives Carl the chalk, puts 1 in his pocket, and waits at the side of the classroom next to Anna.

- At this point it is recommended to pay students' attention to how the stack is being built, where Anna is waiting first and Burt is waiting next to her.
- Carl checks the value in his pocket (*n*) and checks if *1 < 1*. Since this check yields false, Carl continues to the else-clause. He knows how to perform *System.out.println*, and he prints 1 on the blackboard. Then, Carl should perform *mystery(n−1)* and he calls Danna to perform this task. Carl gives Danna the chalk, puts 0 in her pocket, and goes standing at the side of the classroom, next to Anna and Burt.
- Danna looks for the *n*'s value in her pocket and checks if *0 < 1*. Since the condition is satisfied, Danna turns to the then-clause. Since she knows how to perform *System.out.println*, she prints "finished" on the blackboard. At this point, Danna finishes her task, turns to Carl (the little person who asked her to perform her job), returns him the chalk and, after she is thanked by Carl, she returns to her seat.
- In the class, it is recommended to pause the process at this point and ask the students how, in their opinion, the role play should continue. It is also important to emphasize that at this point the 3 little people— Anna, Burt, Carl—are still waiting to continue their jobs from the point they stopped it, but each of them has a different value of n in his or her pockets.
- After Carl thanks Danna for her help, he continues to the last instruction he should perform. He checks *n*'s value in his pocket (sees that it is 1) and prints 1 on the blackboard. Carl turns to Burt (who hired him) and gives him the chalk.
- Burt thanks Carl for his help and continues to the last instruction he should perform. He prints *2* (*n*'s value in his pocket) on the blackboard, turns to Anna and gives her the chalk.
- Anna thanks Burt for his help, prints *3* (*n*'s value in her pocket) on the blackboard, turns to the chief person and returns him the chalk.

- This ends the role play and a whole class discussion about the process starts.
- Stage B: Class discussion
- During the discussion, the students are encouraged to reflect on the process they just saw, addressing its advantages and disadvantages. For example, since the model is based on a metaphor (see Sect. 7.2.6), it is another opportunity to discuss metaphors as a pedagogical tool.
- If time permits, it is recommended to give the students a "reversed" task; that is, the students are given a textual description of the little people role play of another recursive method and they are asked to reconstruct the original method in Java.

12.4.2   The "Top-Down Frames" Model

The little people metaphor is an excellent pedagogical tool for illustrating the execution of recursive methods; individual learners, however, cannot carry it out in their notebook by their own. The following model—the top-down frames model—overcomes this limitation of the little people model by enabling each learner to trace the recursive process by her or his own.

This model guides learners, on each level of the recursion, to write all the instructions that should be executed (one after another), where for each recursive call a new box is opened, until all the boxes are embedded in the initial method invocation (see Fig. 12.1).

Once all the boxes are embedded with all the details, the (printing) instructions can be executed (see Fig. 12.2).

It is important to note that unlike the little people model, the top-down frames model does not emphasize all the issues of the recursive process (the returning process, for example, is not strongly demonstrated here); it is, however, helpful to use this model as a tracing tool in learners' notebooks. It is also important to supply the learners with different models to support their learning of the recursive process.

In the MTCS course, after the instructor explains the top-down frames model, the students are asked to trace by their own another recursive method/program, for example, the following recursive_magic (1, 3) method.

```
public static void recursive_magic (int a, b) {
        if (a==b)
                    System.out.println ("finished");
        else {
                    System.out.println (a);
                    recursive_magic (a + 1, b);
                    System.out.println (a);
        }
}
```

It is recommended to discuss with the students the differences between recursive methods, which do not return a value (i.e., in Java, they return *void*), and recursive functions (i.e., methods that return a value). The recursive examples presented here are methods that do not return a value. It is recommended to demonstrate the two models on a recursive function (e.g., Fibonacci), emphasizing what a little person does with the returned values when he or she finishes his or her job.

Clearly, additional tracing models are presented in the literature. The students can be asked to find additional tracing models by using the web (looking, for example, for tracing demonstrations in YouTube). It is important to emphasize criteria of good models; in the case of tracing recursion, for example, one such criterion is that

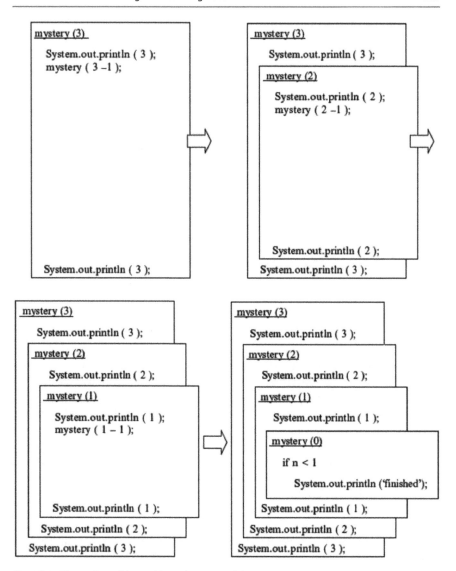

Fig. 12.1 Illustration of the top-down frames model

tracing models should relate to the process invocation, the parameter values, the values of local variables (if exist), and the returning process.

12.5   Research on Learning and Teaching Recursion

This section addresses computer science education research on learning and teaching recursion. As mentioned in Chap. 4, computer science education research may contribute significantly to teachers' knowledge and professional development in at

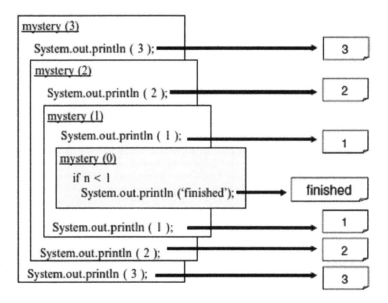

Fig. 12.2 Printing instructions of the top-down frames model illustration

least four ways: (1) becoming members of the computer science education community, (2) increasing teachers' awareness to learners' conceptions and difficulties (strengthen Shulman's (1986) model category of knowledge of learners), (3) strengthening teachers' Pedagogical Content Knowledge (Shulman 1986), and (4) broadening teachers' teaching toolbox.

As Settle (2014) points out "the combination of approaches for teaching recursion and the degree to which students master the topic has generated a significant body of work in the computing education community" (p. 1).

Since recursion is a central computer science concept, as explained at the beginning of this chapter, computer science teachers' familiarity with research on learning and teaching recursion may also contribute to their professional development in each of the above four ways. Specifically, since recursion plays a central role in almost all introductory computer science courses, teachers should be familiar with this educational research area if they wish to become members of the computer science education community; they must be aware of learners' conceptions of recursion and difficulties learners encounter when learning recursion; their pedagogical content knowledge should include examples of recursion and teaching strategies for recursion; and, finally, computer science teachers should broaden their teaching toolbox with respect to recursion. Clearly, these arguments are also applicable for prospective computer science teachers, and therefore, we suggest addressing this research area in the MTCS course. Activities 94 and 95 can be facilitated in the MTCS course to achieve these purposes.

Activity 94: Investigating Research on Learning and Teaching Recursion, Homework

Table 12.3 presents the homework assignment.

Table 12.3 Homework about computer science education research on learning and teaching recursion

Worksheet—research on learning and teaching recursion
Look at the list of research papers on learning and teaching recursion that appears at the bottom of this worksheet.
Choose one paper from the list, read it, and work on the following tasks:
In your opinion, what are the three main messages of the paper?
Indicate three main issues you thought about while reading the paper. You can relate to what you found most interesting, to what increased your curiosity, to issues you would like to read more about, or any other idea you thought about.
Mini-research (optional): Choose one of the papers and reconstruct the research described in the paper.
Paper List
Ford G (1982) A framework for teaching recursion. SIGCSE Bull 14(2):32–39.
Haberman B, Averbuch H (2002). The case of base cases: Why are they so difficult to recognize? Student difficulties with recursion. ITiCSE 2002, pp 84–88.
Haynes S (1995). Explaining recursion to the unsophisticated. SIGCSE Bull 27(3):3–6
Kahney H (1989) What do novice programmers know about recursion? In: Soloway E, Spohrer J (eds) Studying the novice programmer, pp 209–228. Lawrence Erlbaum, Hillsdale
Scholtz T, Sanders I (2010) Mental models of recursion: investigating students' understanding of recursion. ITiCSE 2010, Ankara, Turkey
Tessler J, Bradley B, Calvin L (2013) Using cargo-bot to provide contextualized learning of recursion. Proceedings of the 9th annual international ACM conference on international computing education research (ICER '13), 12–14 Aug 2013, San Diego, San California pp 161–168
Wu C, Dale NB, Bethel LJ (1998) Conceptual models and cognitive styles in teaching recursion. Proceedings of the 30th SIGCSE Technical Symposium on Computer Science Education, Atlanta, GA pp 292–296

Activity 95: Recursive Models, Homework and Presentation in the Course

In order to broaden students' teaching toolbox, in this activity they are asked to find in the computer science education literature models for recursion and to present them in the class (see Table 12.4 for the homework assignment). Another option is to organize a poster session in which each student presents his or her model on a poster.

Table 12.4 Homework about recursion models

Worksheet—models for recursion
Find in the literature at least one model that explains recursion to learners and was not presented in the course (i.e., not the little people metaphor or the top-down frames model). Indicate the paper title, authors, and abstract.
Write a short description of the model and illustrate it on at least one specific recursive program/method.
Discuss the advantages and disadvantages of the model.
Prepare a short (10–15 min) presentation of the model to be presented in the course (or, alternatively: prepare a poster to be presented in the course).

12.6   How Does Recursion Sound?[9]

Section 5.5 (Debugging) and Activity 65 (The musical song debugging activity, Sect. 8.4) aim at increasing learners' awareness to the importance of debugging, debugging processes, and the role of debugging in learning processes. Activity 65 also illustrates the use of the computer lab in computer science education.

In this section, we illustrate how the use of the computer lab can enhance learners' understanding of recursion. For this purpose, we use colors and music for the examination of trees and fractals which are two well-known recursive structures. We suggest that the examination of these recursive structures may foster also learners' general understanding of the concept of recursion.

This approach that fosters the use of colors and sounds relies on the common agreement that, similar to the convention that one picture is worth 1000 words, one musical note is worth at least 100 words. Just like visualization, music is a powerful way to activate people's senses, change people's mental moods, convey messages quickly (e.g., in horror movies), and provide information that sometimes is very difficult, or even impossible, to convey by other means. In this spirit, Vickers (1999) claims that "Although sound is not visible we are still able to construct mental images when presented with particular sounds or pieces of music." (p. 15).

Activity 96 that focuses on fractals and trees is based on three stages. In addition, if the musical debugging activity (see Activity 65 in Sect. 8.4) has not been facilitated yet in the MTCS course, it is recommended to ask students to read the paper on song debugging (Lapidot and Hazzan 2005).

---

[9] Based on Lapidot and Hazzan (2005) © 2005 ACM, Inc. Included here by permission.

Activity 96: Using Colors and Music for the Examination of Recursive Structures

- Stage A: Musical demonstration
- The instructor of the MTCS course executes a program that draws a binary tree. The program plays also a unique tone for each level (e.g., C (Do) for level 1, D (Re) for level 2, and so on). During the first illustration, the students are asked just to watch the program execution. During the second illustration, they are asked to close their eyes, listen to the music, and imagine the drawing of the tree according to the music. The same process can be carried out for a fractal drawing. (See Eglash et al. 2011, for an interesting use of fractal simulations of African design in a high school computing class).
- Stage B: Lab work
- The students work on the activity presented in Table 12.5.

Table 12.5 Student worksheet on recursion in colors

Worksheet—recursion with colors and music
Write a Java program that draws a binary tree.
Color each level of the tree with a different color.
Add music to each level.
Optional: Draw a fruit or a flower at the end of each branch of the tree.
The following picture is a colored binary tree of level 5.

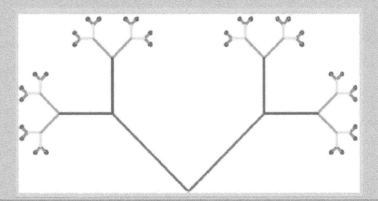

- Stage C: Class discussion
- After the students complete their lab work, it is recommended to facilitate a class discussion that addresses the following issues:
  - What recursive aspects/ideas/perspectives can be learnt based on the tree activity?
  - How can colors and music enhance learners' understanding of recursion, if at all?
  - Why do musical differences between trees and fractals exist? What do these differences tell us about their recursive structure and about recursion in general?

12.7   Assessment

This section includes two activities (97 and 98) related to assessment in the context of recursion to be facilitated in the MTCS course: evaluation of a nonprogramming task and a test construction.

Activity 97: Analysis of Recursive Phenomena
In Chap. 10, we mentioned that a computer science teacher should use a variety of assessment methods. In this spirit, we recommend on a nonprogramming project for the assessment of learners' understanding of recursion, as is illustrated below.

- Stage A: Choosing and analysis of recursive phenomena, homework
- The students work on the worksheet presented in Table 12.6 as a homework assignment.

Table 12.6 Worksheet on recursion in a nonprogramming context

Worksheet—recursion in life		
Choose 2 topics which you find interesting; for example, biology, mathematics, politics, transportation, communication, history.		
For each topic, choose two phenomena that represent different types of recursion. For example, look at the following table:		
	Linear recursion	*Two-dimensional recursion*
Topic 1: food	Phenomenon 1	Phenomenon 2
Topic 2: music	Phenomenon 3	Phenomenon 4
According to the example, phenomena 1 and 2 should have a different recursive nature; the same implies for phenomena 3 and 4.		
Address the four phenomena: Analyze their recursive types, compare them, and discuss connections among them.		
Your homework will be graded as follows: Each phenomenon receives 5 % for its description, and 15 % for its recursive analysis. In addition, 20 % is given for the analysis of the four phenomena, connections among them, and their comparison.		

- Stage B: Class presentation and discussion
- After the students complete their homework, it is recommended to allocate time for the presentation of their work in the class. It is also important to discuss the task both from a cognitive perspective and from a pedagogical perspective. The discussion can be based on the examination of several specific student works and their evaluation. Such a discussion enables to concentrate on the evaluation theme on three levels: (a) the evaluation of this specific task about recursion, (b) the evaluation of a nonprogramming task, and finally (c) general ideas related to evaluation (see also Chap. 10).

Activity 98: Construction of a Written Test on Recursion
Activity 81 in Sect. 10.2 focuses on test construction. Within the context of the current chapter, it is possible to follow all its stages with respect to recursion as the topic of the test. If time does not permit to facilitate all the stages, it is recommended to focus on the test construction stage and to ask the students to construct questions about recursion of different types and difficulty levels (see Chap. 9).

12.8   Additional Activities

In this section, we present additional activities related to learning and teaching recursion to be facilitated in the MTCS course. The purpose of this list is to illustrate again that, indeed, recursion can be the focus of almost any topic discussed in the MTCS course. For the readership convenience, the activities are organized by the order of the chapters in this guide to which they are related.

Activity 99: History of Recursive Functions
Section 3.3 focuses on the history of computer science. An optional activity about recursion in the historical context would ask students to read (e.g., using Internet resources) about the history of recursive functions (such as, Gödel's computable functions and Turing's machine) and to present their findings in the course.

Activity 100: Comparison of Recursive Algorithms in Different Paradigms
Section 3.6 focuses on programming paradigms. In the context of recursion, an optional activity would ask students to compare the same recursive algorithm in different paradigms (e.g., logic, functional, procedural) or compare iterative and recursive solutions to the same problem.

Activity 101: Recursive Patterns
Section 5.4.3 relates to algorithmic patterns. An optional activity for recursion would analyze recursive patterns for list processing, such as map, filter, and reduce (see Table 12.7).

Table 12.7 Recursive patterns for list processing

Recursive patterns for list processing—map, filter, reduce

Map is a function that for a given list L1 and a given unary function F, returns a list L2, in which F was activated on each of L's elements.

Filter is a function that for a given list L1 and a given Boolean function F, returns a list L2 that includes only the elements from L1 that satisfy the F condition.

Reduce is a function that receives a list L1 and a binary function F and returns an object (not necessary a list) that equals the result of the activation of F on all L1's elements.

Examples for recursive patterns

Recursive pattern	Input list L1	Input function F	Output
Map	10, 7, −2, 25	Double	20, 14, −4, 50 (list)
Filter	10, 7, −2, 25	Positive?	10, 7, 25 (list)
Reduce	10, 7, −2, 25	Add	40 (the sum of L1 elements)

Activity 102: Recursion Animation
Chapter 8 focuses on lab-based teaching. In addition to the tree activity presented in this chapter (see Sect. 12.6), it is recommended to use animations (such as the Jeliot environment or a similar IDE) and demonstrate the tracing of Fibonacci or a tree drawing.

Activity 103: The Use of a Tutorial for Exploring Recursive Algorithm Time Efficiency
Chapter 8 focuses on lab-based teaching. The use of tutorials can add several layers to the learning process since each student can progress according to his or her understanding. Pevac (2012) presents the use of an application-based tutor, specifically built to support the learning of recursive algorithms, and reports on the tutor's usefulness in improving student learning. In that context, an optional activity would ask students in the MTCS course to experience the tutor application and comment on its potential usefulness in relation to learning difficulties in understanding of recursion (for example, the difficulties explored in the paper presented in activity 94).

Activity 104: Design of Questions about Recursion
In the spirit of Chap. 9, an optional activity would ask students to build different types of questions for recursion.

Activity 105: Planning the Teaching of Recursion
Chapter 11 examines, among other topics, lesson planning. In the context of recursion, an optional activity would ask students to prepare a lesson, for example, about types of recursion (tail-recursion, double recursion, mutual recursion) or about recursive manipulations of strings. Further discussion can be conducted in relation to the order of integrating the learning of recursion along a curriculum. This discussion can be based for example on the next papers: (1) Bruce, Danyluk, and Murtagh (2005) explain Why they belive that structural recursion should be taught before arrays in CS 1; and (2) Mirolo (2011) challenge the traditions and asks is iteration really easier to master than recursion.

References

Bruce KB, Danyluk A, Murtagh T (2005) Why structural recursion should be taught before arrays in CS 1. SIGCSE Bull 37(1):246–250

Eglash R, Krishnamoorthy M, Sanchez J, Woodbridge A (2011) Fractal simulations of African design in pre-college computing education. ACM Transac Comput Educ (TOCE) 11(3):1–14

George CE (2000) ERSOI—Visualising recursion and discovering new errors. Proceedings of the 31st SIGCSE technical symposium on computer science education, Austin, Texas pp 305–309

Gersting JL (1996) Mathematical structures for computer science (3rd edition). WH Freeman, New York

Götschi T, Sanders I, Galpin V (2003) Mental models of recursion. Proceedings of the 34th SIGCSE technical symposium on computer science education, Reno, Nevada

Harvey B (1997) Computer science logo style—volume 1: Symbolic computing 2/e. MIT Press, Cambridge

Harvey B, Wright M (1999) Simply scheme: introducing computer science 2/e. MIT Press, Cambridge

Hofstadter D (1979) Godel, Escher, Bach—an eternal golden braid. Vintage, New York

Lapidot T, Hazzan O (2005) Song debugging: merging content and pedagogy in computer science education. Inroads—SIGCSE Bull 37(4):79–83

Lapidot T, Levy D, Paz T (2000) Functional programming for high school students. (in Hebrew). Migvan—R & D in Computer Science Teaching, Technion, Haifa

Levy D (2001) Insights and conflicts in discussing recursion: a case study. Comp Sci Educ 11(4):305–322

Mirolo C (2011) Is iteration really easier to master than recursion: an investigation in a functional-first CS1 context. In Proceedings of the 16th annual joint conference on Innovation and technology in computer science education (ITiCSE '11). ACM, New York, p 362

Pevac I (2012) First experiences with tutor for recursive algorithm time efficiency analysis. J Comput Sci Coll 28(1):56–65

Settle A (2014) What's motivation got to do with it? A survey of recursion in the computing education literature. Technical Reports. Paper 23. http://via.library.depaul.edu/tr/23. Accessed 1 June 2014

Shulman LS (1986) Those who understand: knowledge growth in teaching. J Educ Teach 15(2):4–14

Vickers P (1999) CAITLIN: Implementation of a musical program auralization system to study the effects on debugging tasks as performed by novice Pascal programmers. Doctoral thesis, Loughborough University, UK. http://computing.unn.ac.uk/staff/cgpv1/caitlin/index.htm. Accessed 22 Sept 2010

Getting Experience in Computer Science Education

13

Abstract

This chapter deals with the teaching experience that the students enrolled in the Methods of Teaching Computer Science (MTCS) course gain before becoming computer science (CS) teachers. Three frameworks in which the prospective CS teachers gain their first teaching experience are presented: (1) The practicum, which takes place in high school, after one or two semesters of learning the MTCS course; (2) CS teacher training within the Professional Development School (PDS) collaboration framework; and (3) a tutoring framework that can be integrated in the MTCS course. We also present activities that can be facilitated in the MTCS course, in which the students deal with and analyze teaching scenarios taken from the practicum of other prospective CS teachers.

13.1 Introduction

This chapter deals with the teaching experiences that the students gain before becoming CS teachers, in which they implement what they have learned in the Methods of Teaching Computer Science (MTCS) course. The importance of these first teaching experiences stems from the recognition that one significant way to acquire pedagogical-disciplinary knowledge involves activities performed in actual teaching situations. These activities provide opportunities and guide the teacher toward reflective processes that address learners' thinking (Wilson and Berne 1999; Putnam and Borko 2000).

We present three frameworks in which the students gain this first teaching experience: The practicum, which takes place in high school, after one or two semesters of learning the MTCS course; a practicum model within a collaborative framework defined as Professional Development School (PDS); and a tutoring framework that can be integrated in the MTCS course. In addition, we present activities that can be

© Springer-Verlag London Limited 2014

O. Hazzan et al., *Guide to Teaching Computer Science*,
DOI 10.1007/978-1-4471-6630-6_13

facilitated in the MTCS course, in which the students deal with and analyze teaching scenarios taken from the practicum of other prospective CS teachers.

The teaching experiences in a variety of environments increase students' confidence as CS teachers, as well as elevate their awareness to students' learning processes. Therefore, some institutions offer in addition special laboratory teaching or micro-teaching courses. In these courses, prospective teachers practice a variety of teaching situations in a friendly environment (to a small group of pupils or peers) with a close guidance of an instructor. If such courses are not available, and the MTCS remains the only opportunity in which the prospective CS teachers can gain any teaching experience, then the instructor of the MTCS course should try to find additional opportunities to let the students experience and reflect on CS teaching situations before practicing real teaching situations in the schools. For example, and as has already illustrated in this Guide, short presentations can be integrated in the MTCS course, in which the students teach their peers a CS topic or present to their peers the product of an activity they worked on.

13.2   The Practicum in the High School[1]

13.2.1   General Description

This section focuses on one of the central components of CS teacher preparation programs—the practicum—the stage in which the prospective CS teachers practice real CS teaching situations in high schools. The objective of the in-school teaching practicum is to bring prospective teachers closer to the field work of teachers while actually teaching the knowledge domain (Eick et al. 2004). With respect to high school CS teaching, the importance attributed to the practicum is expressed, for example, in the Model Curriculum for K-12 Computer Science, prepared by the ACM K-12 Task Force Curriculum Committee (Tucker et al. 2004), which outlines standards that refer to the preparation of CS teachers (see also Stephenson et al. 2005).

The practicum is carried out in different ways. Some programs require a full year participation in school activities; others require that the practicum be performed for a specific, shorter period of time. In all these cases, however, as has been mentioned above, the main objective of the practicum is to let the prospective teachers experience what real teaching is, before becoming CS teachers.

To achieve this goal, the practicum is usually performed with the guidance of two CS educators: an in-school mentor, a CS high school teacher who trains the student and guides him or her during the practicum; and a university mentor who is a faculty member in charge of the academic aspects of the practicum. In most cases, the academic faculty member who teaches the MTCS course is also the one who accompanies the students in their practicum. During the period in which the students are in the school, they accompany their in-school mentor, observe lessons taught by him or her, assist in various activities, and, of course, at a certain stage, start

---

[1] Based on Hazzan and Lapidot (2004), © 2004 ACM, Inc. Included here by permission.

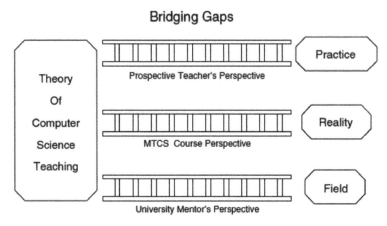

Fig. 13.1 Theory—practice/reality/field gaps. *MTCS* Methods of Teaching Computer Science

teaching themselves (in the broader sense, including lesson preparation, teaching in the class or the computer lab, preparing and grading exams, etc.).

The involvement of the university mentor is usually expressed by periodic visits to the schools in order to observe lessons taught by the prospective CS teachers. Reflection and feedback meetings take place after each such lesson. Thus, the university mentor continues the guidance started in the MTCS course. In this sense, the practicum can be viewed as one of the teaching methods employed in the MTCS course.

Usually, the students are asked to submit a report (a kind of reflection) about their experience in the school. Sometimes, they are asked to carry out additional activities such as, conducting some research, attending a workshop at the university in parallel to the practicum, or participating in school faculty meetings.

13.2.2 The Practicum as a Bridge Between Theory and Its Application

Hazzan and Lapidot (2004) examine the practicum through three lenses: the prospective CS teacher's standpoint, the MTCS course viewpoint, and the university mentor's perspective. For each perspective they highlight the importance of the practicum by explaining how it helps bridge a specific gap related to the theory of CS teaching (see Fig. 13.1): a gap between theory and practice (the prospective CS teacher's perspective), a gap between theory and reality (the MTCS course perspective), and a gap between theory and the field (the university mentor's perspective). These three perspectives are explained in what follows.

13.2.2.1 Prospective Computer Science Teachers' Perspective: Bridging the Gap between Theory and Practice

This perspective refers to the following questions: Why cannot we let the students start teaching their own classes immediately after they finish studying the MTCS course? Why are the different activities carried out in the MTCS course, such as the discussion of different teaching approaches and micro-teaching, insufficient? The answer to these questions is derived from several reasons.

First, teaching is an apprenticeship profession, such as Medicine. This implies that part of the students' professional preparation should include experience in the *real* environment in which they will teach in the future, that is, high school CS classes. In other words, an appropriate preparation toward CS teaching in the high school should include practice in the high school with a close guidance of an expert; this approach is clearly different than that of letting the students start teaching in real high school CS classes just after the MTCS course, before gaining a proper practice.

Second, the answers to the above questions are also based on the active-learning teaching approach (see Chap. 2); that is, the practicum provides the prospective CS teachers with significant experience in real high school classes that none of the situations integrated either in micro-teaching courses or in the MTCS course, afford. For example, during the practicum, the students may feel the need to use different teaching methods, to which they were exposed in the MTCS course (see Chap. 7), for teaching of different CS topics. That is, though many different teaching methods are discussed in the MTCS course, only their *actual* implementation in real teaching situations, together with a reflection process that follows it, can improve the students' understanding with respect to the essence of CS teaching.

Third, the practicum is yet another opportunity in which the students can improve their understanding of *CS* concepts. This improvement happens while they prepare the lesson to be taught in the high school, while they are teaching the lesson, and finally, in the reflection session that takes place, either with the in-school mentor or university mentor, after each lesson the student teaches in the school. Clearly, it is important to gain this improved understanding prior to becoming a CS teacher. Activity 106 below, to be facilitated in the MTCS course, illustrates this knowledge construction related to CS.

Finally, from an organizational perspective, entering the school environment, as an organization, is not a simple task. One has to become familiar with the school culture, procedures, roles, behavior styles, professional language, and more. The practicum provides the prospective CS teachers an opportunity to be exposed to the organizational aspect of the school prior to becoming a member of the community (either of the school he or she does the practicum in or of another school); this preliminary familiarity may ease their entrance to the school as CS teachers.

Accordingly, the above reasoning delivers the message that the practicum constitutes a significant stage in the construction process of the prospective CS teachers' professional perception which also reduces *gaps between theory and practice*. In other words, the practicum can help the students close gaps between the *theory* they learn in the MTCS course and in other pedagogical courses and the actual *practice* of CS teaching. Furthermore, it is important to increase the students' awareness to these gaps as well as to different ways to bridge them.

13.2.2.2   MTCS Course's Perspective: Bridging the Gap Between Theory and Reality

The MTCS course and the practicum are both important components of CS teacher preparation. While each one alone is not sufficient for the CS teachers' training, they do mutually contribute to each other. Accordingly, from the MTCS course's perspective, the practicum bridges a *gap between theory and reality*.

As we can see in Activities 106–108, the presentations of scenarios that took place during the practicum in the MTCS course may help bridge the *gap* between the *theory* that is taught in the MTCS course and *reality*—what actually goes on in schools. For example, based on a real lesson taught by a prospective CS teacher during the practicum, a detailed description of the lesson can be presented to the students participating in the MTCS course. The students are asked to analyze the lesson, to point out good teaching behavior, to suggest alternative actions for specific teaching behaviors, to analyze pupils' answers, etc. The fact that the description is based on an actual lesson taught by a prospective CS teacher, who has similar teaching experience to that of the students in the MTCS course, is important since it may serve the students as a kind of self-examination of their own actions.[2] Such real case studies can be collected by the university mentor or taken from research done by other scholars in the field (see Chap. 4).

From a broader perspective, the MTCS course should not be based solely on theory. Had the MTCS course been based only on theory and hypothetical case studies, the gap described in the previous section from the students' prospective would have been further widened and, consequently, their entry into the practicum would have been even a more mysterious and difficult process.

13.2.2.3   University Mentor's Perspective: Bridging the Gap Between Theory and the Field

This section discusses the gap that exists between the university environment (in which the university mentor is active) and the high school environment (in which CS is taught), that when examined from the university mentor's perspective, may be bridged by the practicum. By discussing the bridging of this gap, we actually also explain why it is important that a university mentor be part of the practicum.

First, since in some cases the university mentor teaches the MTCS course, his or her involvement in the practicum can create continuity between the MTCS course and the practicum. Specifically, ideas that are presented and discussed in the MTCS course can be referred to first, in the ongoing guidance that the university mentor gives the prospective teachers during the practicum, and second, in the reflection meetings that take place after lessons observed by the university mentor. We highlight that one of the main aims of the reflective meetings is to educate the students to become a reflective practitioner (Schön 1983, 1987; see also Chap. 5). Further, by eliciting reflective processes in these meetings, the university mentor can actually help the prospective CS teachers bridge the gap between theory and practice (discussed above).

Second, university mentors do not usually teach in a school, yet they prepare their students to become CS teachers. Thus, their involvement in the practicum provides them with an opportunity to be involved in the field that is the subject of the MTCS course and to avoid the well-known gap between the academia and the high school system. Thus, the *gap* between the *academia* and the *school* is bridged.

---

[2] We note that the analysis of teaching situations taken from lessons taught by experienced teachers is important for other purposes.

Third, each visit in a high school is also a rich source for new material to be addressed in the MTCS course. Therefore, the university mentor should increase his or her awareness to events, scenarios, interaction styles, and typical behaviors that can be brought back to the MTCS course, and by doing so closes also the *gap* between *theory* and *reality* mentioned above.

Finally, for the success of the practicum process, it is important that the in-school mentor and the university mentor have direct communication channel and good relationship. The university mentor's visits in the school can support building such relationships.

Activities 106–108 should be facilitated in the MTCS course just before the students start their practicum. They are based on the analysis of real scenario descriptions, taken from the practicum of prospective CS teachers in the high school.

It is recommended, however, that the instructor of the MTCS course brings authentic examples from lessons he or she observed as a university mentor while mentoring prospective CS teachers in their practicum.

Activity 106: Bridging Gaps Related to the Content Aspect of Computer Science Education

• Stage A: Scenario description[3]

The students are presented with the following scenario, in which Anna, a prospective CS teacher, was asked by her in-school mentor to prepare a 2 h lesson about procedures to a 11th grade students, who learned CS in the procedural programming paradigm. In a meeting with her university mentor that took place prior to the lesson Anna said:

Maya (Anna's in-school mentor) asked me to prepare a lesson about procedures. It will be the first time that this class learns procedures. I thought about it a lot at home and it seems to me that I don't have what to teach with respect to this topic to fill 2 h. It is a simple topic and it is sufficient to give them [the pupils] the syntax of how to write procedures. Even if I present many examples, I will not have what to do for the entire period of time. So, I thought that during the second hour I would start to teach them about types of parameters.

Anna was encouraged by the university mentor to explain what it means "to know (or to understand) procedures," that is, what procedures literacy is according to her understanding.

In the MTCS, it is recommended that the instructor stops and asks the students how they define "procedures literacy."

Anna listed the following issues:

---

[3] Based on Lapidot (2005).

- A procedure is a subprogram.
- A procedure helps us think simpler, because each procedure is a small task, a subtask, and if the task is complicated, it is divided into subtasks.
- We should understand the syntax of procedures, how they are called and where in the program they are written.
- A procedure within a procedure: Indeed, it is not always necessary, but such an option exists.
- Oh! There are too many questions that I haven't thought about before.
- There is also recursion.

After that conversation, Anna taught the lesson. During the lesson she addressed the topic of procedures from different perspectives. After the lesson, in the reflection meeting, Anna said:

At the end, I had about 15 min left and I still had to talk about what a procedure gives us in general—hierarchy. But I did not prepare myself for this and I had no idea how to do it. It is not a simple task to explain why we need hierarchy and even now I do not know how to find an example for this topic. In general, it is funny that at the beginning I thought that there is nothing to do with procedures, I thought that a quarter of an hour would be sufficient [to teach procedures] and that we would then continue with types of parameters.

- Stage B: Scenario analysis, work in pairs and a discussion

After the scenario is presented to the students (either orally or in writing), they are asked to work in pairs on the following questions:

1. What is the source of Anna's initial conception that the notion of procedure can be taught in about 15 min?
2. Can you imagine a scenario that took place during the lesson that Anna taught which gave her a hint that more time is needed for the teaching of the notion of procedure?
3. What questions, in your opinion, were raised by the high school pupils in that lesson? For each question, describe what would be your answer as a teacher.
4. Which additional aspects of the procedure concept, in your opinion, should be addressed while teaching this topic?
5. Suggest an example that illustrates the notion of hierarchy, for which it was difficult for Anna to find an appropriate example in the lesson.

After the students work on these questions, a discussion takes place in which their suggestions are discussed. In this discussion, it is important to address both pedagogical aspects (mainly, teaching methods and class management issues) and cognitive topics (mainly, students' understandings and (mis)conceptions). Question 3, for example, in which the students are asked to envision what actually happened in the lesson, highlights these two perspectives and at the same time increases the students' awareness to learners' perceptions (see Chap. 6).

After these questions are discussed, this case can be summarized by highlighting the following two topics which are clearly illustrated by Anna's case:

- Technical teaching (which emphasizes technical aspects of CS) versus conceptual teaching (which encompasses also nontechnical CS issues):
  - As could be seen, the need to teach procedures led Anna, guided by the university mentor's help, to add conceptual topics to the technical picture she drew first, such as the contribution of procedures to problem-solving situations.
  - It is highly relevant to discuss with the students what pupils learn from each mode of teaching and what kind of tasks (see Chap. 9) are appropriate for conceptual teaching (vs. technical teaching).
- Challenges involved in teaching CS soft ideas (such as, a procedure) to high school CS pupils (see Sect. 3.7).

We note that a gap was also reduced from the university mentor's perspective. Specifically, it is reasonable to assume that Anna's case increased the university mentor's awareness to the fact that what is taught in the MTCS course is not transferred automatically to in-school situations and that the issue of technical versus conceptual teaching should be further emphasized in the MTCS course.

Activity 107: Bridging Gaps Related to the Pedagogical Aspect of Computer Science Education

- Stage A: Scenario description

The students are presented with the following scenario, in which Jim taught the topic of Bubble Sort to 10th grade pupils, managing the lesson very successfully. Yet, Jim did not encourage any pupil–pupil dialogue and the entire class interaction was based on teacher–pupil discourse. More specifically, Jim did not ask the pupils, even once, to answer a question asked by another pupil or to respond to an answer presented by other pupils. He was the only one who addressed any particular idea suggested by the pupils.

When this issue was presented to him in the reflection meeting that took place after the lesson, he could easily reflect on his class management style. As it turns out, Jim was aware of interaction-related issues. He explained, however, that he had based the lesson on his own interaction with the pupils because he wanted to follow his lesson plan. As it turned out, he even did not consider the option of achieving the lesson objectives by incorporating in the lesson also pupils–pupil interactions.

- Stage B: Scenario analysis, class discussion and summary

This class discussion focuses on the characteristics of Jim's behavior. It is recommended to highlight the importance of the practicum as an arena in which the students can gain and improve also pedagogical skills, and in particular, increase their awareness to the option of fostering also pupil–pupil interaction to improve teaching processes.

Specifically, during the discussion and at its summary, it is important to highlight the following behaviors which are typical to new teachers and to discuss with the students ways to cope with these tendencies:

- The desire to follow their lesson plan.
- Avoidance of pupil–pupil interaction. Several reasons may explain this avoidance:
  - The wish to follow lesson plans
  - The lack of confidence required to manage the complex situation of guiding a full class interaction
  - Inability to recognize the added value they, as well as the pupils, can gain by following up on pupils' assertions

If appropriate, this discussion can also address feelings associated with teaching in general, and CS teaching, especially in the first years, in particular.

We end by mentioning that similar to the scenario described in Activity 106, Jim's case can increase the university mentor's awareness to the fact that what is taught in the MTCS course is not transferred automatically to in-school situations. Consequently, when the university mentor guides prospective CS teachers, he or she should increase their awareness to the importance of using different interaction modes (e.g., teacher–pupils, pupil–pupil).

Activity 108: Prospective Teacher's Conception about the First Lesson

This activity focuses on the first lesson a novice teacher teaches.

The students are presented with the following scenario about Gila, who is a prospective CS teacher who conducted her practicum in an 11th grade CS class. After the first lesson she taught, Gila confessed:

Yesterday, at home, I told myself maybe 100 times what I should cover [in the lesson]. In the morning, it was no longer important for me what I will say in class. On the bus [on the way to the school] I started talking to myself again in my head. I realized that I know every single word and I was afraid that I will use other words. So, I stopped talking to myself, but still remembered the lesson like a song.

When the university mentor asked her:

What exactly did you say in your head?

Gila answered:

I had no idea what they [the pupils] will say so I couldn't think about them. Since not all learners face the same difficulties and since I was familiar only with the difficulties I faced [as a learner], I could concentrate only on these difficulties. Therefore, I could think only about what I will say in the class.

Then, Gila stated:

On the very first lesson someone teaches, it's not important what he [or she] will say. The only important issue is that the teacher will not be afraid and just talk about something. [...] **The first lesson that a new teacher teaches does not contribute much to his pupils' learning; it is more important that s/he himself [the teacher] will learn.**

As in Activities 106 and 107, following the scenario presentation, a discussion can take place in which the following topics, as well as others, may be addressed:

1. Gila's last statement: Do the students in the MTCS course agree with this statement? Do they disagree?
2. The preparation process of the first lesson that a novice teacher teaches.
3. Novice teachers' feeling before and after the first lesson they teach.

13.3 Computer Science Teacher Training Within the Professional Development School Collaboration Framework[4]

13.3.1 General Description and Main Objectives

The PDS is a collaboration framework aims to assist teachers in taking their first steps in their professional career. The rationale is to form innovative foundations through partnerships between schools and professional education programs within academic institutions (Clark 1999). The collaborative effort aims to improve the initial preparation of prospective teachers in parallel to the enhancement of the professional development of in-service classroom teachers. A PDS is a learning-centered community, and its partners are guided by a common vision of teaching and learning that is grounded in research and practitioner knowledge. The objectives of the PDS within a specific school are determined jointly by all partners: schools ad-

---

[4] Based on Ragonis and Oster-Lewintz (2011), © 2011 ACM, Inc. Included here by permission.

ministrations, subject-matter teachers, and the teacher preparation institute, and are designed to advance the mutual interests of all partners. The building of a learning community is based on the sharing of disciplinary and pedagogical knowledge, reflective processes, research on the teaching of the discipline, bridging theory and practice, construction of teaching-learning activities, planning long-term projects, and encouraging initiative implementation (Teitel 2003). According to the PDS approach, the collaboration enhances change and mutual development in both systems (schools and academic institutions) and minimizes the gap between them (Korthagen and Kessels 1999; Levine 2003). In particular, this framework empowers the prospective teachers and gives them a broader and deeper base for their future work as teachers.

The objectives of the student–teachers' practicum within the PDS in a specific school are defined on three levels (Klieger and Oster-Levinz 2008):

1. *Social-institutional collaboration:* exposing the prospective teachers to the school as an active institution and involving them in the school activities. Examples of such activities are: facilitating a volunteer project, coping with challenges of special-needs pupils, and collaborating with teachers in their routine tasks during recesses throughout the school day.
2. *Collaboration in the teaching field:* actively involving the prospective teachers in various teaching activities, such as: establishing and running online learning programs at the school, guiding pupils' research and projects, providing weaker pupils with individual assistance, integrating student-teachers into teaching, and into exam writing, assessing, and marking.
3. *Professional advancement:* developing a professional learning community in the school that comprises student-teachers, pedagogical supervisors, and mentor-teachers. In general, this level is considered to be the most important one since it cannot be easily achieved by other frameworks in which prospective teachers gain experience with the profession of teaching.

13.3.2   Training Computer Science Prospective Teachers Within the PDS

Each student-teacher practices at a specific school for 1 day a week throughout the entire academic year. At the beginning of the academic year, a meeting is held between the student-teacher, the school mentor-teacher, and the academic pedagogical supervisor with the objective of constructing a schedule for the student-teacher's practicum day at the school. Each student has an individual daily schedule that is flexible and can be changed during the year according to specific needs that emerge and ask to include various experiences.

A typical practicum day at the school consists of a combination of the following activities:

Observing Lessons   The student-teachers observe lessons that the mentor teacher's teaches in the different classes, including lab lessons.

Table 13.1 Example of a student teacher's day of practicum

Lesson	Activity
1	Observing a grade 12 computer lab lesson given by the mentor teacher
2	Observing a grade 12 lesson on binary trees given by the mentor teacher
3	Meeting with the mentor-teacher and the pedagogical supervisor analyzing the lesson on binary trees
4	CS staff meeting on composing a grade 10 test on If statements
5	Meeting with the pedagogical supervisor as a preparation for teaching a lesson on two-dimensional arrays in grade 11 that will take place in the following week
6	Meeting with the pedagogical supervisor and all CS student-teachers in preparation for the tutoring activity
7	Tutoring grade 10 pupil. The main topic: If statements

Teaching Lessons The student-teachers teach lessons and gradually expand their level and length. At the beginning of the year, they teach only part of a lesson; for example, they may explain part of a program to the class. Later on, they may teach a full problem-solving process, gradually progressing to teaching an entire lesson, for example, on nested If statements. By the end of the year they teach an entire teaching unit, for example, arrays.

Active Participation The students-teachers are active participants in the CS staff meetings (the learning community), in which various teaching issues are addressed, for example, the issue of integrating visualizations.

Development and Grading Tests The student-teachers develop tests (for example, the final test on loops) together with the mentor-teachers and the pedagogical supervisors, and help the mentor-teachers grade the pupils' tests.

Supporting Pupils with Special Needs The student-teachers help pupils with special needs during tests, for example by reading the tests and dictating them.

Fulfilling Teacher Obligations Student-teachers are also involved in other daily teachers' obligations, such as, monitoring pupils in the school yard during the recess, a duty that all teachers must fulfill.

Personal Meetings Student-teachers hold personal meetings with their mentor-teachers and pedagogical supervisor.

Group Discussions The student-teachers, as a group, discuss with other student-teachers from the same discipline or from another disciplines various teaching and learning situations, successful events, and difficulties and challenges they face while teaching.

Tutoring Pupils Student-teachers tutor pupils in their problem-solving processes, guide them and help them overcome their obstacles.

Student-teachers' schedules are flexible so that they have the opportunity to practice the various activities throughout the entire year. Table 13.1 presents an example of a student-teacher's practicum schedule at the beginning of the school year.

In addition to the presented various activities, the student are also responsible for leading special initiatives in the school. Such initiatives are usually not directly related to CS, but school officials consider the CS students teachers and their pedagogical supervisor as part of the professional information and communication technologies (ICT) community. For example, student teachers may lead an Online Day in which all students stay at home and participate in online activities in various disciplines. The CS student teachers help the school teachers develop the online activities and help them activate the activities throughout the Online Day. Another example of an ICT school activity is running a simulation program for computerized elections, which student teachers can develop during a national election year and which the school can use on Election Day to simulate an election process.

13.3.3   The Practice of Teaching Within the PDS

When a student teaches a lesson, all the other students in the course are invited to observe it. A feedback discussion is conducted at the end of each lesson, led by the academic pedagogical supervisor. The mentor-teacher and the other CS prospective teachers, who observed the lesson, participate in this discussion. The feedback session starts with the student who taught the lesson; he or she addresses his or her feelings during the lesson, and reflects on the lesson plan compared with its implementation as well as on what in his or her opinion was done properly and what can be improved. The student-teacher leads and directs the discussion to issues he or she deems important.

In addition to the individual practicum days, the annual practicum plan includes two full weeks of practice at the school, one in each of the two semesters, during which there are no lectures at the academic institution and the student-teachers spend the entire week at the schools. This is a unique opportunity to experience the continuity of teacher's work. Such a full week of practicum enriches all PDS partners. In particular, during this week, the CS prospective teachers teach in different classes, something they cannot experience during a single practicum day, enabling them to become acquainted with and experience various learning methods. They also have the opportunity to participate in homeroom classes and in other school events and activities.

More specifically, the goals of the week-long practicum (as opposed to single days) are: (a) to expand the students opportunities for acquaintance with the school system, including all of its strata and activity settings, which they were unable to experience during their 1-day-a-week practicum; (b) to enable the students to teach a sequence of lessons; (c) to give students an opportunity to authentically experience the teacher's work during an entire week, which helps them understand the essence and complexity of teacher's work; (d) to enable a meaningful dialogue with mentor-teachers, school staff, other students, and the pedagogical supervisor in order to fully understand the underlying aspects of various educational processes; and (e) to enhance informal relationships among students, mentor-teachers, and high school pupils, and to practice how to develop relationship with their future teaching partners.

During the 2 weeks of practicum, the prospective teachers also join their pedagogical supervisors to visits in various high-tech companies and in other high schools that implement other CS curricula (if exist) or have a special, state-of-the-arts computer lab or other unique features and equipment. In general, the objective of all these activities is to expand, as much as possible, the student-teachers' school experience, as well as their CS knowledge and pedagogical perspective.

13.4   A Tutoring Model for Guiding Problems-Solving Processes[5]

In this section we present a tutoring model whose objective is to develop and establish the pedagogical-disciplinary knowledge of prospective CS teachers with respect to guiding learners in problem-solving processes.

The tutoring model focuses on the *tutor*, who is a prospective CS teacher (i.e., a student) enrolled in the MTCS course, and is based on hands-on teaching experience. It comprises individual tutoring whereby the students, whose CS knowledge is more established, tutor novice CS students enrolled in an introductory CS course. In other words, during the tutoring process, tutor–students, the prospective CS teachers learning the MTCS course, who have already acquired CS knowledge, support the learning processes of novice CS learners. Since the mentoring model is integrated in the MTCS course, it is actually based on an active application of pedagogical-disciplinary knowledge acquired in the MTCS course in actual teaching situations.

In what follows we focus on the actual implementation of the tutoring model. Additional details can be found in Ragonis and Hazzan (2008, 2009a, 2009b).

13.4.1   The Implementation of the Tutoring Model

Tutoring takes place in one or two series, each of which includes five sessions with a single tutee—a student (or high school pupil) who is taking an introductory CS course. Focus is placed on problem-solving processes and the activity is accompanied by guided reflective processes. These teaching situations enable the tutors to cope with learners' difficulties in understanding different CS concepts and in problem-solving processes. The model, thus, implements principles of constructiv-

---

[5] Based on Ragonis and Hazzan (2009a). Copyright 2009 by the Association for the Advancement of Computing in Education (AACE). [http://www.aace.org] Included here by permission.

Table 13.2 Tutoring session feedback worksheet

Tutoring session feedback worksheet
A. General
Describe the topic of the session
Describe the problem discussed
Describe the course of the session
B. Tutor feedback
What concept/s do you think constituted a difficulty for the tutee?
Describe the tutee's difficulty/misunderstanding/misconception/…
What teaching tools did you use to help the tutee overcome the difficulty/misunderstanding/ misconception?
Did you use knowledge that you acquired in the Methods of Teaching Computer Science course or in another course? Specify what knowledge you used
What more would have helped you give the necessary assistance? (additional disciplinary knowledge, additional teaching knowledge, what knowledge, which tools? …)
If you could repeat this tutoring session, what would you do differently? Why?
What is your personal feedback at this stage of the tutoring? (what is the nature of the communication between you and your tutee? the quality of support? do you feel you are advancing the tutee student? are you benefiting from the tutoring? are there difficulties? …)

ist teaching in the context of CS education (see Ben-Ari 2001 and Chap. 2) as well as principles of situated learning (Lave and Wenger 1991; Stein 1998).

Specifically, during the sessions, the tutee raises problems and the tutor guides the tutee through the problem-solving process. Tutoring is based on the identification of learner difficulties and the subsequent application of different teaching strategies to overcome such difficulties. The serial nature of the sessions enables the tutor to receive feedback on the knowledge the tutee acquired in previous sessions, and thus, in fact, to receive feedback on his or her own teaching process. Reflective processes are integrated into the process; at the end of each session, the tutor is required to complete a feedback sheet that guides him or her to rethink the session and focus on the teaching objectives and on the teaching methods applied.

The mentoring process is accompanied by a tutoring coordinator, who coordinates the process and provides the students with ongoing support throughout the entire tutoring period in the form of a coaching process. The coordination and guidance of the tutoring model can be considered to be meta-tutoring, since the coordinator guides the tutors in order to advance them while learning teaching skills; in other words, the tutor–student are the coordinator's tutees. At the same time, the tutor–students tutor their tutees and, here too, their objective is to promote the tutees' learning. Similar to the role of the university mentor in the practicum, it is preferable that the coordinator of the mentoring process would be the instructor of the MTCS course.

In addition to the coordinator of the mentoring process, the support mechanism of the tutoring model includes (a) an introductory meeting at the beginning of the MTCS course in which the tutoring model is explained, and (b) online support forums for discussing tutoring sessions and posing questions on disciplinary-related topics.

In more detail, the tutor requirements are:

- To identify a tutee from among students enrolled in an introductory CS course
- To hold five tutoring sessions, each lasting about 2 h
- To complete a feedback sheet for each tutoring session (see Table 13.2) and to submit it to the tutoring coordinator
- To hold two individual meetings with the tutoring coordinator: one, following the first tutoring session and the second, after completing the series of tutoring sessions
- To present the MTCS course plenum with one episode from the tutoring process
- To complete a final summarizing feedback questionnaire
- The mentoring model has several essential guidelines for its implementation:
- It is important that the five tutoring sessions of each tutor are held with the same tutee. Such a relationship enables continuity of activity, reference to previous sessions, and development along with the learning material. Continuity enables the tutor to see the impact of the sessions on the tutee and to examine, for instance, what still has not been properly understood and which thinking strategies the tutee has adopted.
- A 10-h tutoring framework (5 sessions of 2 h each) seems reasonable in terms of the tasks required of a student in the MTCS course. It is recommended holding face-to-face meetings, which can be combined with electronic communications according to the tutee's needs.
- Since the prospective CS teachers will teach, in the future, in high schools, tutoring a high school pupil would seem to have been more appropriate in terms of experiencing the true target audience. Nevertheless, the need to find a high school pupil might constitute a problematic constraint for the students, since their living and learning environment is the campus, in which both they and tutee are present. If a student has access to a tutee who is a high school pupil, it can be approved. In some cases, a group of up to three tutees can be tutored together.
- Completing a feedback questionnaire after each tutoring session is essential to the tutors' learning process. Focused and reflective examination of their actions during the session enables tutors to evaluate their performance and formulate guidelines for themselves for the remaining sessions.

13.4.2   The Contribution of the Mentoring Model to Prospective Computer Science Teachers Teaching Experience

The mentoring model has the potential to foster the skills of the prospective CS teachers on three levels:

1. Promoting *the pedagogical-disciplinary* professional skills by means of identifying learners' difficulties in real-life situations, assisting and facilitating learners

in overcoming their difficulties, adopting a teacher–researcher perspective, and developing a relationship between tutors through a process of creating copartnership in a learning community.

2. Promoting the *pedagogical* professional skills by encouraging the students to reflect on their teaching, fostering a teaching approach that develops the learners' thinking and developing guidance tools that include the formation of interpersonal relationships with learners combined with the implementation of teaching methods that suit the learners.

3. Promoting the *disciplinary* knowledge as a by-product of the guidance process. Coping with others' difficulties enhances nuances in the understanding of disciplinary concepts perhaps not encountered by the prospective CS teachers as learners, in the spirit of the well-known slogan "Teaching is the best way to learn."

A research conducted on one specific application of the mentoring model (Ragonis and Hazzan 2009a) found that during the mentoring process the prospective CS teachers:

- Became aware of the importance of identifying learners' difficulties
- Emphasized problem-solving processes
- Became aware of the need to adapt their teaching process to different learners
- Adopted reflective thinking processes and encouraged these processes among their tutees as well (see also Ragonis and Hazzan 2010)
- Reinforced their own self-confidence regarding their ability and place in the disciplinary teaching process
- Realized the contribution of the tutoring model to their training as future CS teachers

13.5 Practicum Versus Tutoring

Though the purposes of the practicum in schools and the mentoring experience are similar, that is, to provide the students with an opportunity to gain some teaching experience before becoming high school CS teachers, these two teaching experiences are different. We mention three differences between the two teaching experiences.

First, the responsibility of the teaching process is different in the two cases. While in the mentoring process, the prospective CS teachers are the responsible figures on the entire teaching process, the practicum in the school is limited to a small number of lessons taught by the students, and in most cases they are treated as guests, even in cases in which more profound models of co-teaching (Eick et al. 2004) or of PDS take place (see Sect. 13.3 and Darling-Hammond 2001; Furlong 2000; Teitel 2003). In these frameworks, the main responsibility of teaching the discipline does not lie with the prospective teacher, but with the regular class teacher.

Activity 109: Prospective Teacher's Conception about Computer Science Teaching in an Informal Framework

This activity focuses on CS education in an informal framework, that is, not in school settings and not necessary to young learners.

The students are asked to locate an informal teaching framework in which CS is taught, to visit one lesson in this framework and to teach one lesson.

Based on this experience they are asked to reflect on similarities and differences between this framework and that of teaching CS in the schools.

Second, when teaching a class, the prospective teachers usually have concerns about the degree of cooperation they will receive from the learners as well as other problems involving class management and discipline. These concerns do not usually enable the students to experience two essential pedagogical concepts: one, follow-up on each learner's learning processes of the knowledge domain and, two, the impact of their teaching methods on each learner. Needless to say that sensitive prospective teachers may pay attention to each learner's progress, and further, effective guidance of the in-school mentor and the university mentor should address pupils' learning processes and problem-solving processes. These learning processes, however, are more transparent in a one-on-one mentoring process.

Finally, one clear difference between the practicum and the mentoring model is the easiness of their facilitation. That is, the mentoring model can be facilitated in the university, without the need to coordinate it with the high school administration and with a high school teacher. Therefore, when practicum in the high school is not an available framework, it is recommended to let the students enrolled in the MTCS course gain some teaching experience in the framework of a mentoring process.

References

Ben-Ari M (2001) Constructivism in computer science education. J of Comput Mat Sci Teach 20(1):45–74

Clark RW (1999) Effective professional development schools: agenda for education in a democracy. Jossey-Bass, San Francisco

Darling-Hammond L (2001) When conceptions collide: constructing a community of inquiry for teacher education in British Columbia. J Educ Teach 27(1):7–21

Eick CJ, Ware FN, Jones MT (2004) Coteaching in a secondary science methods course: learning through a coteaching model that supports early teacher practice. J Sci Teach Educ 15(3):197–209

Furlong J (2000) School mentors and university tutors: lessons from the english experiment. J Theory Pract 39(1):12–19

Hazzan O, Lapidot T (2004) The practicum in computer science education: bridging gaps between theoretical knowledge and actual performance. ACM SIGCSE Bull 35(4):29–34

Klieger A, Oster-Levinz A (2008) In search of the essence of a good school: school characteristics leading to successful pds collaboration. Aust J Teach Educ 33(4):40–54

Korthagen FA, Kessels JPM (1999) Linking theory and practice: changing the pedagogy of teacher education. Educ Res 28(4):4–17

Lapidot T (2005) Computer Science teachers' learning during their everyday work. Unpublished Ph. D. Thesis, The Department of Education in Technology and Science, Technion-Israel Institute of Technology

Lave J, Wenger E (1991) Situated learning: legitimate peripheral participation. Cambridge University Press, Cambridge

Levine M (2003) Foreward. In: Teitel L (ed) The professional development schools handbook: starting, sustaining and assessing partnerships that improve student learning. Corwin Inc, Thousand Oaks, pp XIII–XVII

Putnam RT, Borko H (2000) What do new views of knowledge and thinking have to say about research on teacher learning? Educ Res 29(1):4–15

Ragonis N, Hazzan O (2008) Tutoring model for promoting teaching skills of Computer Science prospective teachers. 13th Ann. Conf. on Innov. and Technol. In Comput. Sci. Educ.-ITiCSE, Madrid, Spain, pp 276–280

Ragonis N, Hazzan O (2009a) Integrating a tutoring model into the training of prospective computer science teachers. J Comput Math Sci Teach (JCMST) 28(3):309–339

Ragonis N, Hazzan O (2009b) A tutoring model for promoting the pedagogical-disciplinary skills of prospective teachers. Mentor Tutor: Partnersh Learn 17(1):67–82

Ragonis N, Hazzan O (2010) A reflective practitioner's perspective on computer science teacher preparation. Proc. 4th ISSEP, Zürich, Switzerland, pp 90–106

Ragonis N, Oster-Levinz A (2011) Pre-service computer science teacher training within the professional development school (PDS) collaboration framework. In Proc. of the 5th Int. Conf. on Informatics in Schools: Situation, Evolution and Perspectives (ISSEP'11), Ivan Kalaš and Roland T. Mittermeir (Eds). Springer-Verlag, Berlin, Heidelberg, pp 106–116

Schön DA (1983) The reflective practitioner. BasicBooks, New York

Schön DA (1987) Educating the reflective practitioner: towards a new design for teaching and learning in the profession. Jossey-Bass, San Francisco

Stein D (1998) Situated learning in adult education. ERIC Digest #195. http://ericacve.org/docs/situated195.htm. Accessed 30 Jan 2007

Stephenson C, Gal-Ezer J, Haberman B, Verno A (2005) The new educational imperative: i high school computer science education, Final report of the CSTA, Curriculum Improvement Task Force, http://csta.acm.org/Communications/sub/DocsPresentationFiles/White_Paper07_06.pdf. Accessed 14 July 2010

Teitel L (2003) The professional development schools handbook: starting, sustaining and partnerships that improve student learning. Corwin Inc, Thousand Oaks

Tucker A, Deek F, Jones J, McCowan D, Stephenson C, Verno A (2004) A model curriculum for K-12 Computer Science: Report of the ACM K-12 Education Task Force Computer Science Curriculum Committee, Assoc. for Comput. Mach., New-York

Wilson SM, Berne J (1999) Teacher learning and the acquisition of professional knowledge. Rev Res Educ 24:173–209

Design of Methods of Teaching Computer Science Course

14

Abstract

This chapter describes how to design a Methods of Teaching Computer Science (MTCS) course within an academic computer science teacher preparation program, and suggests two possible syllabi for such a course. It is emphasized, however, that different approaches and frameworks can be applied when one designs the course. In the first section of this chapter, we propose four possible perspectives on the MTCS course: the NCATE standards, merging computer science with pedagogy, Shulman's model of teachers' knowledge, and research findings. The second section of the chapter describes two MTCS course syllabi. We mention that though the focus in this section is placed on the MTCS course, the course models, as well as parts of them, can be used also for other purposes related to computer science teaching, such as curriculum design and professional development of computer science teachers.

14.1 Perspectives on the MTCS Course[1]

As could be observed so far in this Guide, the education and training of prospective computer science teachers, in general, and the MTCS course, in particular, should address a broad spectrum of topics. Accordingly, when designing the MTCS course, one can use different approaches for course organization.

In this section, we propose four possible perspectives on the MTCS course that can inspire the course design according to the instructor's preferences: (1) the NCATE standards, (2) merging computer science with pedagogy, (3) Shulman's model of teachers' knowledge, and (4) research findings. In the following description of the four perspectives, the readers can observe that many of the presented topics have already been addressed in this Guide.

---

[1] Based on Lapidot and Hazzan (2003).

© Springer-Verlag London Limited 2014
O. Hazzan et al., *Guide to Teaching Computer Science*,
DOI 10.1007/978-1-4471-6630-6_14

1. *The NCATE standards* (Tucker et al. 2003): Among these 13 standards, teachers are expected to learn how to plan lessons/modules related to programming process and concepts, to be able to develop assessment strategies appropriate to lesson goals (Standard 3), and address student population characteristics (Standard 4). From this perspective, the MTCS course can be organized by addressing the different standards.

2. *Merging computer science concepts with pedagogy*: This perspective on the MTCS course is based on the special amalgam of the two disciplines: computer science and pedagogy. Although pedagogical principles and teaching methods are learnt in other general pedagogical courses of teacher preparation programs, the MTCS course should focus on their implications and adoption into the context of computer science education. Accordingly, this perspective highlights topics such as the introduction and summary of a specific topic, learning in groups, learning by inquiry, and planning constructivist activities, all in the context of computer science education. It also addresses the uniqueness of the following topics in relation to computer science topics: creative and nonconventional use of the computer laboratory; analysis of teaching difficulties and possible obstacles; adjustment of learning materials for learners with different needs; definitions and their role in learning processes; and the use of metaphors, multimedia, and games in computer science education.

3. *Shulman's model of teachers' knowledge*: This perspective on the MTCS course is inspired by Shulman's model of teacher knowledge base (Shulman 1987), which consists of seven categories: content knowledge, general pedagogical knowledge, curriculum knowledge, pedagogical content knowledge, knowledge of learners and their characteristics, knowledge of educational context, and knowledge of educational targets. Though it is relevant to include all these categories in the MTCS course, following Shulman's recommendation (1987, p. 20), we propose that the pedagogical content knowledge (PCK) category is the most important one. From this perspective, we mention a few core concepts that may be included in the MTCS course with respect to their PCK: programming and algorithm design, data representation and information organization, debugging, and soft ideas. Instructors of the MTCS course should address such topics in the course, with full attention to a variety of aspects, such as: analogies, illustrations, examples, explanations, types of questions, and learners' difficulties.

4. *Research findings*: This perspective is based on research findings. The extensive research in computer science education conducted especially in the last two decades can highlight known misconceptions or difficulties and recommended teaching strategies. A variable is an example for such a topic. Variables play an essential role in most programming languages and, as indicated by computer science education researchers, learners often face difficulties in understanding the various aspects of the subject.

14.2 Suggestions for MTCS Course Syllabi[2]

This section describes two syllabi for the MTCS course and partially illustrates an implementation of the above perspectives on the MTCS course.

We reemphasize that the proposed course structures and lessons are only two options that the instructor of the MTCS course can apply and that, in fact, the course can be designed in many different ways. In practice, one should integrate activities into the MTCS course to meet the needs of the prospective teachers and the local computer science curriculum. For this purpose, it is recommended to use a table (such as Table 1.1) in the design process of the course.

Thought the focus in this section is placed on the MTCS course, the course models, as well as parts of them, can also be used for other purposes by all those engaged in computer science teaching: curriculum developers, lecturers in teacher training programs, computer science instructors, and lecturers involved in the professional development of computer science teachers.

14.2.1 Course Structure

The course presented here consists of 112 h of classes and training, divided into two semesters. Each week there are two 2-h-long lessons. Each of the two semesters is devoted to different high school curriculum parts. Specifically, in the illustrative course, the first semester focuses on the teaching of foundations of computer science, whereas the second semester focuses on the teaching of more advanced topics, such as abstract data types and computational models. The course lessons, however, can be adapted for the teaching of any high school computer science curriculum.

14.2.2 Course Syllabus

In what follows, we present in detail two syllabi of the course taught at the first semester and give only a brief description of the course taught at the second semester. We note that this Guide includes more contents needed for one specific MTCS course, and instructors of the MTCS course who design the course should take this fact into consideration. Activity 110 suggests several options how to start the first lesson of the MTCS.

---

[2] Based on Ragonis and Hazzan (2008) Disciplinary-pedagogical teacher preparation for pre-service Computer Science teachers: Rationale and implementation, Informatics in Secondary Schools—Evolution and Perspective—ISSEP 2008, Lecture Notes in Computer Science, Vol. 5090/2008: 253–264. Included with permission here.

Activity 110: First Lesson of the MTCS Course

This activity outlines several options on how to start the first lesson of the MTCS.

- Ask the students to suggest the main topics that should be included in a high school computer science curriculum.
- If a state curriculum exists, let the students become familiar with it.
- Ask the students what a computer science teacher should know.
- Present the question: In what sense does the teaching of computer science differ from the teaching of other disciplines?
- Ask the students to share one learning episode they experienced in their learning of computer science.
- Ask the students to design the first lesson in their future high school computer science class.
- Delve into one of the topics presented in this Guide. For example, facilitate the students with the activity on the lab-first approach presented in Sect. 8.3. Such an experience enables the students to experience the active learning model and taste, in the first lesson of the course, different topics that will be elaborated later in the course.

14.2.2.1   First Semester—Fundamentals of Computer Science—Option 1

- Lesson 1: Introduction to the high school computer science curriculum
- Lesson 2: What is computer science?
- Lesson 3: Research in science education
- Lesson 4: Reflection and its application as a teaching and learning tool
- Lesson 5: Programming paradigms
- Lesson 6: The fundamentals of teaching object-oriented programming
- Lesson 7: Demonstration of different Java development environments
- Lesson 8: Teaching planning of a computer science topic—The case of variables
- Lesson 9: Types of questions
- Lesson 10: Teaching conditional expressions and statements
- Lesson 11: Integrating the Internet into computer science teaching
- Lesson 12: Teaching loops statements
- Lesson 13: Participating in the National Conference for Computer Science Teachers
- Lesson 14: Integrating the computer lab into the teaching process
- Lesson 15: Development and analysis of algorithms
- Lesson 16: Issues of teaching memory organization
- Lesson 17: The object-first approach for teaching introduction to computer science
- Lesson 18: Diversity in computer science education

- Lesson 19: The history of computer science
- Lesson 20: How to write an exam?
- Lesson 21–22: How to evaluate an exam?
- Lessons 23–24: Guiding software projects development
- Lessons 25–28: The final semester work and peer teaching

  - *The final semester work*: See Activity 90 in Chap. 11.
  - *Peer teaching*: The last semester lessons are dedicated to individual, 30-min presentations of the final semester work given by each of the students. Presentations include four parts:
  - A summary of the research paper dealing with the study unit.
  - Introduction of a short excerpt from the planned lesson.
  - Teaching the excerpt to the classmates.
  - Description of the considerations involved in the development of this segment of the lesson.

14.2.2.2 First Semester—Fundamentals of Computer Science—Option 2

- Lesson 1–2: Introduction to computer science education: Karel the robot, the lab first approach, visualization-and animation-based IDEs
- Lesson 3: Introduction to the high school computer science curriculum
- Lesson 4: Variables—metaphors, preparing an exhibition poster, concept map construction
- Lesson 5: Variables—how to start the teaching of variables; planning the teaching of variables
- Lesson 6: Variables—prepare a lab on variables
- Lesson 7: Variables—Computer science education research on variables
- Lesson 8: Learners' alternative conceptions
- Lesson 9–10: Types of questions and test construction
- Lesson 11: Control structures—Pedagogical examination of a classification activity
- Lesson 12: Control structures—Pedagogical examination of games
- Lesson 13: Pedagogical examination of rich tasks
- Lesson 14–15: Mid-semester summary: Overview of teaching methods and class organization methods
- Lesson 16–17: Song debugging, using music in computer science teaching, teaching soft ideas, reflection and its application as a teaching and learning tool
- Lesson 18: Teaching arrays
- Lesson 19: Development and analysis of algorithms, algorithmic patterns
- Lesson 20: Teaching the object-oriented programming paradigm
- Lesson 21–22: Programming paradigms
- Lesson 23: Guiding software projects development
- Lesson 24: Pedagogical examination of the CS-Unplugged approach
- Lesson 25: Integrating the Internet into computer science teaching

- Lesson 26–27: Getting experience in computer science education
- Lesson 28: The nature of computer science

14.2.2.3   Second Semester—Teaching Advanced Topics in Computer Science: Abstract Data Types and Computational Models

The second semester deals with the teaching of advanced computer science contents. Lessons held during this semester are related to the following subjects:

- The teaching of abstract data types (list, stack, queue, and binary tree);
- The teaching of computational models (deterministic finite automata, push-down automata, Turing machine, languages);
- The teaching of advanced disciplinary concepts such as recursion, complexity, and abstraction;
- Integration of social aspects, e.g., ethics, in the teaching process.

References

Lapidot T, Hazzan O (2003) Methods of teaching computer science course for prospective teachers. Inroads SIGCSE Bull 35(4):29–34

Ragonis N, Hazzan O (2008) Disciplinary-pedagogical teacher preparation for pre-service Computer Science teachers: rational and implementation, informatics in secondary schools—evolution and perspective—ISSEP. Lect Notes Comput Sci 5090:253–264

Shulman LS (1987) Knowledge and teaching: foundations of the new reform. Harv Educ Rev 57(1):1–22

Tucker A, Deek F, Jones J, McCowan D, Stephenson C, Verno A (2003) A model curriculum for K-12 Computer Science. Final Report of the ACM K-12 Activity Force Curriculum Committee. http://csta.acm.org/Curriculum/sub/CurrFiles/K-12ModelCurr2ndEd.pdf. Accessed 17 Aug 2014

High School Computer Science Teacher Preparation Programs

15

Abstract

This chapter puts the Methods of Teaching Computer Science (MTCS) course in the wider context of computer science teacher preparation programs. It first describes a model for high school computer science education that one of its components is computer science teacher preparation programs. The model consists of five key elements—a well-defined curriculum, a requirement of a mandatory formal computer science teaching license, teacher preparation programs, national center for computer science teachers, and research in computer science education—as well as interconnections between these elements. Then, the focus is placed on the teacher preparation programs component of the model, describing (1) a workshop targeted at computer scientists and computer science curriculum developers who wish to launch a computer science teacher preparation programs at their universities but lack knowledge about the actual construction of such programs and (2) the perspective that examines computer science teaching as an additional profession for computer science graduates.

15.1 A Model for High School Computer Science Education[1]

This section presents a model for high school computer science education. It is based on an analysis of the structure of the Israeli system of high school computer science education. The model consists of five key elements as well as interconnections between these elements.

---

[1] Based on Hazzan et al. (2008) © 2008 ACM, Inc. Included here by permission.

© Springer-Verlag London Limited 2014
O. Hazzan et al., *Guide to Teaching Computer Science*,
DOI 10.1007/978-1-4471-6630-6_15

15.1.1   Background

In the final report of the Association for Computing Machinery (ACM), K–12 Task Force Curriculum Committee (Tucker et al. 2003), the Israeli high school computer science curriculum has been mentioned to illustrate the fact that "the development of K–12 computer science is making more headway internationally than in the United States." The report continues: "In Israel, a secondary school computer science curriculum (Gal-Ezer and Harel 1999) was approved by the Ministry of Higher Education and implemented in 1998. It blends conceptual and applied topics, and is offered in grades 10, 11, and 12" (p. 6). In 2010, the curriculum has been updated in light of the new developments in the field of computer science. The curriculum comes in two versions: a basic version and an extended one.

It is proposed, however, that the Israeli high school computer science *curriculum* is not the only contributor to the Israeli system of computer science education in the high school. Based on an analysis of this system, a model for high school computer science education is sketched (Hazzan et al. 2008). Here, we describe an updated version of this model.

The model consists of interrelationships among five key components:

- A well-defined curriculum (including written course text books and teaching guides)
- A requirement of a mandatory formal computer science teaching license
- Teacher preparation programs (including at least a Bachelors degree in computer science and a computer science teaching certificate study program)
- National center for computer science teachers
- Research in computer science education

It is proposed that each of these components, as well as the relationships among them, establishes the solid infrastructure of the Israeli high school computer science program, strengthens it, and makes it, as is indicated by the ACM K-12 Task Force Curriculum Committee report, one of the leading computer science high school curricula in the world.

The next section presents the details of the model. The description is partially based on Hazzan et al. (2008).

15.1.2   The Model Components and Their Amalgamation

Figure 15.1 presents the model which reflects the structure of the Israeli system of high school computer science education.

In what follows we first elaborate on the five components of the model. Then connections among these components are described.

National High School Computer Science Curriculum Here are several of the key principles that guided the curriculum development (Gal-Ezer et al. 1995):

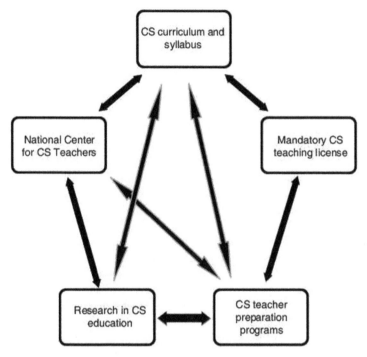

Fig. 15.1 A model for high school computer science education

- Computer science is a full-fledged scientific subject. It should be taught in high school on par with other scientific subjects.
- The program concentrates on the key concepts and foundations of the field.
- Each of the two versions of the curriculum has mandatory units and electives.
- Conceptual and experimental issues are interwoven throughout the curriculum.
- A well-equipped and well-maintained computer laboratory is mandatory.
- New course text books and teaching guides must be written for all parts of the program.
- Teachers certified to teach the subject must have adequate formal computer science education. An undergraduate degree in computer science is a mandatory requirement, as is a formal teachers' certificate program of studies.

In addition, the national curriculum describes the rationale of high school computer science education and lays out the topics taught in each unit of the program. Text books and teaching guides are provided for each unit. The teaching guides specify pedagogical aspects of the given topics, such as guidelines for designers of teaching and learning material, recommended lesson plans, solutions for selected exercises presented in the pupils' text books, additional problems to offer to the pupils, plausible learners' difficulties, and additional clarifications related to the learning material.

Mandatory Computer Science Teachers' License In Israel, in order to teach computer science in high school, a teacher should have both a Bachelors degree in computer science and a teaching license in computer science, or, alternatively, a B.Ed. in computer science education. Only then will he/she be authorized by the Ministry of Education to teach computer science in high schools[2]. The computer science teaching diploma can be achieved in two different frameworks:

a. Programs such as the ones described in the "Teacher preparation programs and inservice training" component of the proposed model (see below).
b. Specific training programs, offered to computer science graduates without a teaching license and to teachers of another scientific topic who wish to switch to computer science education. These programs are usually offered after school hours, or in school vacations.

The importance of this component of the model is acknowledged also by the Computer Science Teachers Association (CSTA), which in its 2013 "Bugs in the System: Computer Science Teacher Certification in the U.S." report declares: "This report reveals that it is difficult to draw broad conclusions about the certification of computer science teachers in the country beyond the fact that it is not working. Each state has its own process, its own definition of Computer Science, and its own ideas about where it fits in a young person's educational program (if at all). The report and what it reveals about these processes form the basis for a number of policy recommendations:

Establish a system of certification/licensure that ensures that all Computer Science teachers have appropriate knowledge of and are prepared to teach the discipline content."

Teacher Preparation Programs and in-Service Training In most cases, teacher preparation programs are taught in universities or colleges. The prospective computer science teachers study for a Bachelors degree in computer science while at the same time take teacher preparation program courses (which are equivalent to one academic year) during the 4 years of study. The contents of these programs correlate with the statement that "beyond the mastery of core computer science material, good computer science educators should also be familiar with a significant body of material that will expand their perspectives on the field, and consequently, enhance the quality of their teaching" (Gal-Ezer and Harel 1998).

A typical teacher preparation program includes general pedagogical courses (such as, psychology and educational philosophy), basic teaching skills, and specific courses about computer science teaching.

---

[2] In fact, a Masters degree is required for teaching any subject in the Israeli high school system. It is, however, difficult to meet these requirements and in most cases high school teachers only have a Bachelors degree. Unfortunately, and as it happens in other countries, some teachers, mainly those who joined the system many years ago, do not have even a Bachelors degree in computer science. They usually have a Bachelors degree in another scientific subject and switched to teaching of computer science for different, administrative as well as personal, reasons.

The two main courses which deal specifically with the teaching of computer science are the Methods of Teaching Computer Science (MTCS) in the high school course (whose teaching this guide is dedicated to) and the practicum in high school computer science classes (see Chap. 13). One important principle of these courses is the links between theory and practice by highlighting their field component (see also Computer Science Teaching Methodology Course piloted during the spring quarter of 2012 in the Center X Teacher Education program at UCLA[3]).

We mention that the computer science teacher preparation programs serve also in-service computer science teachers by offering them ongoing training about the curriculum, teaching methods and developments in the field of computer science. Indeed, it is reasonable to assume that it is easier to organize and facilitate trainings for in-service computer science teachers within the existing infrastructure of the computer science teacher preparation programs.

National Center for Computer Science Teachers[4]   In 2000, the Israeli Ministry of Education established a National Center for Computer Science Teachers. This center is considered as the professional home for all Israeli computer science teachers. It serves as a bridge between the computer science teachers and the academic disciplinary knowledge and research. These bridges are constructed via conferences, workshops, seminars and courses, a website[5], and a computer science education journal.

The center activities are organized around five major themes:

- Helping create a professional community of computer science teachers
- Fostering the professional leadership of computer science teachers
- Supporting, assisting, and consulting academic computer science education groups and computer science teacher educators and researchers.
- Collecting and distributing computer science education knowledge and experience.
- Researching and evaluating computer science teachers' needs and the center's activities.

The above themes reflect the essence of the center as the professional home of computer science teachers which enables them to be updated in an on-going manner with respect to the frequent curricula development and changes, which especially characterize the discipline of computer science. In addition, we should remember that computer science teachers usually work in schools alone or in small teams and therefore, the teacher center helps them be connected to the larger community of computer science teachers.

---

[3] Computer Science Teaching Methodology Course: http://csta.acm.org/Curriculum/sub/CurrFiles/CSMethodologyCourseSyllabus.pdf

[4] Based on Israeli National Center for Computer Science Teachers (2002). © 2002 ACM, Inc. Included here by permission.

[5] See http://cse.proj.ac.il/index-en.htm

Research in Computer Science Education Intensive research in computer science education is carried out by Israeli researchers who are usually involved in the development of the text books and the teaching guides of the nationwide curriculum. In many cases, this research is carried out during the development process of the material. The purpose of this research is to guide the development process of the text books so that the final product fits high school computer science pupils' level, and it will be possible to teach it during the given period of time.

A typical research project that accompanies the development of new material usually involves both the development team and a group of teachers who agree to be the first to try the new material and to participate in research activities conducted with respect to that material. The research activities mainly include reflective talks about the teaching process, discussions about pupils' conception of the learned topics, interviews with pupils, and observations within the pioneer classes.

Such a research process is iterative; at each step, lessons are learnt and are implemented in the next edition of the developed material. Also, during the first stages of the development process, the research focuses on small number of teachers and classes; as the development of the material proceeds, additional teachers join the group of teachers who teach the new developed material. Along such a cyclic process, the teacher and pupil populations who use the developed material gradually increase, and the text books and teaching guides are shaped and converged toward its final form.

As a bonus, this process boosts the computer science education research in Israel beyond just research directly connected to the evaluation of the course text books and teaching guides. This additional research addresses, for example, undergraduate students' understanding of computer science concepts (cf. Aharoni 2000; Armoni et al. 2006; Dubinsky and Hazzan 2005; Sherman-Kolker 2009). It is plausible to assume that the infrastructure needed for this additional research (expertise in education research as well as in computer science itself) has been established and shaped by the need to evaluate the high school material during its development process.

15.1.3 Connections Among the Model Components

Connections exist among the five elements of the proposed model for high school computer science education. We outline most of them.

Mandatory Computer Science Teaching License and Teacher Preparation Programs Clearly, a mandatory computer science teacher license requires teacher preparation programs to be established to grant these certificates; on the other hand, when teacher preparation programs exist, it is just natural to establish and demand a computer science teaching license to exploit the experience and knowledge shared by these programs.

Teacher Preparation Programs and the National Center for Computer Science Teachers The National Center for Computer Science Teachers serves both in-service and prospective computer science teachers. For both in-service and prospective computer science teachers, the center's website and the journal it publishes include a rich information hub of pedagogical material. Specifically, for in-service teachers, the center provides a life-long learning environment that supports their professional development.

Teacher Preparation Programs and Research in Computer Science Education Teacher preparation programs should include some research elements, such as reading assignments of papers which deal with computer science education research, and mini-research projects, carried out by the students themselves. Thus, instructors of computer science teacher preparation programs should be familiar with research in computer science education, and hence the community of practitioners who are interested in computer science education research enlarges. On the other hand, when the community of computer science education researchers is well established, it wishes to deliver its achievements to target audience. One appropriate framework for this purpose is teacher preparation programs. In addition, research work are carried out on topics related to teacher preparation programs and to in-service teachers training (e.g., Brandes et al. 2010), which further tighten the interconnections between these two components of the model.

Research in Computer Science Education and High School Computer Science Curriculum and Syllabus The existence of a computer science education research community, with its accumulative experience, eases and guides the development process of the text books and the teaching guides that accompany the curriculum. Further, from an organizational nationwide perspective, the mere existence of a research infrastructure enables decision makers to promote the development of the curriculum and to allocate the needed resources for this purpose. This, naturally, boosts the curriculum development. On the other hand, as described above, it is critical to accompany the development of a high school computer science curriculum and syllabus with research that assesses the implementation of the curriculum and evaluates the developed material. A few examples of such research works are presented in Armoni and Gal-Ezer (2003), Gal-Ezer and Zur (2004), and Levy (2000).

The National Center for Computer Science Teachers and Research in Computer Science Education Some of the center's activities serve as a research field for computer science education researchers (see, e.g., Brandes et al. 2010; Haberman et al. 2003; Kolikant and Pollack 2004; Lapidot and Aharoni 2008; Ragonis and Haberman 2003). The center also introduces the computer science teachers with themes in computer science education research. Specifically, several literature reviews were prepared for teachers on topics such as novice difficulties, recursion learning and teaching, and pedagogical patterns.

15.1.4   Comments About the Model

The rapid growth of the computing filed highlights the importance of computer science education. The above model is based on the Israeli experience with respect to a nationwide high school computer science curriculum. Clearly, it is interesting to ascertain the model's applicability to other places on the globe. Further, since the model consists of five components which are interrelated to each other, it is not always clear which component is the chicken and which is the egg. In other words, if a state wishes to apply the Israeli model, how should it proceed? One possible answer for this question is that the model implementation can be initiated by the establishment of computer science teacher preparation programs. The next section describes one plausible way to foster the creation of such programs.

15.2   Construction of a Computer Science Teacher Preparation Program—the ECSTPP Workshop[6]

In this section, we present the rationale, structure, and contents of a proposed workshop on the Establishment of a Computer Science Teacher Preparation Program (ECSTPP), targeted at computer scientists and computer science curriculum developers who wish to launch a computer science teacher preparation program at their universities but lack knowledge about the actual construction of such programs.

15.2.1   Rationale

Based on the central position of computer science teacher preparation programs in the presented model in Sect. 15.1, it is suggested that countries/states that wish to adopt the model may consider starting its application with the facilitation of the proposed ECSTPP workshop, which eventually will foster the establishment of computer science teacher preparation programs.

15.2.2   Population

The ECSTPP workshop is designed for the following groups of computer science practitioners:

- Computer scientists who wish to establish a computer science teacher preparation program in their universities, but are not familiar with the practice of teaching computer science in the high school and with the research in computer science education.

---

[6] Based on Hazzan et al. (2010).

- Designers of high school computer science curricula who lack the background in computer science education research. It is important that these practitioners participate in the workshop since any university that wishes to establish a computer science teacher preparation program is likely to recruit them to teach some of the courses in the program.

15.2.3 Objectives

The workshop's objectives are derived from the specific populations for which it is designed. Specifically, the workshop participants:

- Become familiar with a typical structure of computer science teacher preparation programs
- Become familiar with the research in computer science education
- Become reflective practitioners as educators of high school computer science teachers

15.2.4 Structure and Contents

The proposed ECSTPP workshop comprises three consecutive stages: the Common Ground stage, a 3-day seminar, and the Action stage. The Common Ground stage and the Action stage take place at the participants' institutions before and after the seminar, respectively.

15.2.4.1 Stage 1: Common Ground

The Common Ground stage has two main purposes: first, to prepare the ECSTPP workshop participants for the seminar by increasing their awareness to meaningful themes in high school computer science education, from both the pupil's and the teacher's perspectives; second, to create a common knowledge basis for the workshop participants on high school computer science education, which will serve as the basis for the seminar.

To achieve these purposes, the ECSTPP workshop participants are asked to complete the following assignments prior to their arrival at the seminar:

- Become familiar with their national high school computer science curriculum (if such exists) or with another high school computer science curriculum on which they intend to base their computer science teacher preparation program.
- Spend at least 6 h in high school computer science classes (specifically, at least 3 consecutive hours in two classes) and summarize their observations and insights.
- Write a reflective essay about their own acquisition of computer science concepts during their professional development.

15.2.4.2 Stage 2: Three-Day Seminar

The proposed seminar consists of twelve 1.5-h sessions. It is suggested to schedule it for three consecutive days (e.g., 1st day—4 sessions; 2nd day—5 sessions; 3rd day—3 sessions). Other scheduling options are also possible of course.

The seminar contents are organized in four layers:

Layer 1—Introduction This layer addresses the rationale and structure of computer science teacher preparation programs.

Layer 2—The Methods of Teaching Computer Science (MTCS) Course The MTCS course is one of the central components of any computer science teacher preparation program. As has been explained several times in this Guide, in the MTCS course, the prospective computer science teachers become familiar with the pedagogical context knowledge (PCK; Shulman 1986) of computer science teaching.

Layer 3—High School Practicum The practicum is also a vital component of any computer science teacher preparation program (see Chap. 13). During the practicum, prospective computer science teachers get their first experience in high school teaching.

Layer 4—Computer Science Education (CSE) Research This layer includes an overview of the research in computer science education, discussions of specific computer science education research works, familiarity with common computer science education research methods, and a preliminary experience in computer science education research.

Table 15.1 presents the proposed topic of each session, as well as the layers to which it belongs. As can be seen, the layers are intertwined throughout the seminar in order to highlight their interrelations. For additional details about the contents of each session, see Hazzan et al. (2010).

15.2.4.3 Stage 3: Action

The last session of the seminar (Session #12) is dedicated to the launching of the Action stage which begins right after the seminar ends. The purpose of this stage is to guide the participants of the ECSTPP workshop in the actual establishment of computer science teacher preparation programs in their respective universities.

For this purpose, the participants of the ECSTPP workshop are offered the following two activities:

- Conduct two kinds of research:
- Mini-research in high school computer science classes in order to improve the understanding of the setting of high school computer science teaching.
- Action research (Lewin 1948) about their own process of constructing a computer science teacher preparation program.
- Participate in an online forum with other ECSTPP workshop participants to maintain the spirit of the learning community created during the seminar itself.

Table 15.1 The schedule of the ECSTPP seminar

#	Topic	Layer
1	Gathering, introduction and creating a community of learners	1—Introduction
2	Structure of a computer science teacher preparation program	1—Introduction
3	The MTCS course—Part 1	2—MTCS course
4	Introduction to research in computer science education	4—CSE research
5	The MTCS course—Part 2	2—MTCS course
6	The MTCS course—Part 3	2—MTCS course
7	Research methods in computer science education	4—CSE research
8	The practicum—Part 1	3—Practicum
9	A reflective practitioner's perspective of computer science education	4—CSE research
10	The practicum—Part 2	3—Practicum
11	Action research	4—CSE research
12	Launching the action stage of the workshop	Integration of the four layers

The forum enables participants to share their experience, to learn from each other's experience, and to discuss problems they encounter during the establishment of the computer science teacher preparation program in their own universities.

15.2.5 ECSTPP Workshop—Summary

The ECSTPP Workshop focuses on the construction of computer science teacher preparation programs. The purpose of the workshop is to enable its participants to return to their respective institutions with the basic knowledge needed to start this construction process. Such a workshop has the potential to help initiate the construction of computer science teacher preparation programs that, in turn, according to the model presented in Sect. 15.1, may foster the creation of the needed infrastructure for high school computer science education on a national level.

15.3 Computer Science Teaching as an Additional Profession

This section presents a new approach towards computer science teaching that proposes a solution to three social phenomena: (a) The shortage in science, technology, engineering, and mathematics (STEM) teachers, especially on the high school level, that many western countries experience, (b) people desire to contribute to the society; and finally, (c) the structure of the work force which enables professionals, who wish to change their career, to start a new one at the age of 40+.

In what follows we briefly describe a program that implements this approach (more details can be found in Hazzan and Ragonis 2014). The program is offered by a university that has a STEM teacher preparation program and it provides its graduates an additional profession—high school STEM teachers—that they will be able to use if, when and where they choose to switch to education. In what follows we focus on the computer science graduates who chose to study in the computer science education track.

15.3.1   Program Description and Rationale

The program invites all the university's graduates back to the university to study toward an additional bachelor's degree in one of a STEM education study tracks, one of which is the computer science education track. The graduates who enroll in the program receive full study scholarships from the university for 2 years. Since the number of credits required to complete this degree is similar to that which is required for an MBA, the study program is organized similarly: The students attend classes 1 full day or 2 half-days a week for 2 years, and can continue working as scientists/engineers in the industry in parallel to their studies.

Although the students enrolled in the program receive full study scholarships for two years, they are not required to commit to teaching in the education system. This decision was made for the following reasons:

1. Teaching should not be done only because one made a commitment to do so.
2. The knowledge gained in the program—mainly learning and teaching processes—is useful also in the hi-tech industry when coping with new knowledge and technological developments. Thus, even if the students decide not to switch to education, they will still use the knowledge they gained in their studies and contribute to the country's prosperity, but in a different way.
3. The exposure of students during their studies to schools, to high school students, and to teaching and learning processes will continue to affect them throughout their entire life. Thus, even if they did not originally consider becoming STEM high school teachers, they may consider this option at some future time, either in parallel to their work in the industry or if they ever decide to switch careers to STEM teaching.
4. Those who work in the hi-tech industry can contribute to the educational system in parallel to their work in different ways: part time job (e.g., teaching one class), writing learning material, mentoring high school pupils who develop projects, and more.

15.3.2   The Computer Science Education Track of the Program

The computer science education track of the program is open to students and graduates of the faculties of Computer Science, Information Systems, and Computer Engineering. The program consists of 36 credits as listed below:

- Twenty-two credits of required courses: Introduction to Developmental Psychology, Introduction to Social Psychology, Introduction to Cognitive Psychology, Teaching Methods and Skills, Philosophy of Education, Methods of Teaching CS I and II, School Practicum in Computer Science, Reflective Workshop on Teaching Practicum, and Selected Problems in Computer Science
- Eight credits of elective educational courses
- Six credits of elective advanced courses from the faculty of computer science (seminars, projects, and advanced studies)

15.3.3 Computer Science Students' Perspective and Contribution

Here are several selected reasons for joining the program mentioned by the computer science alumni, who joined the program:

- Have always dreamt about going back to study in the academia and the program was an appropriate opportunity to realize that dream
- Have dropped out of the race for promotion and now have the time to replenish the soul
- Are considering a professional alternative with more conventional work hours
- Are engaged in instruction and thought that the program would contribute to their work

The students study alongside the university's regular undergraduate students, a fact that affects the regular courses studied in the program. Among the feelings these graduates expressed about their studies, they mentioned that they:

- Very much enjoy returning to academia, studying and connecting to current contents as well as to basic computer science contents
- Enjoy the encounter with younger students
- Are challenged by the hands-on experience in the schools and in the mentorship activities

In addition to the alumni's fulfillment of their expectations, teaching in the computer science teacher preparation program also benefits from this program, as is described below:

- The two populations—undergraduate students and alumni—benefit from each other: The alumni, bring their work experience into the class, while the undergraduate students help them relearn and refresh computer science contents that they learned (sometimes many) years ago.
- The undergraduate students' self-esteem increases since they realize that they chose a profession that successful computer science graduates wish to study as a second career.
- The diversity inherent in this cohort of students further enhances the abovementioned phenomena.

Finally, we mention that since schools are going to undergo various changes in the near future, due to technology and other cultural factors, these graduates in general, and the computer science graduates in particular, will be well equipped to lead the process, since:

- They can introduce the schools to a different organizational culture, one they have experienced in the hi-tech industry, including that of start-ups and international markets. This culture is innovation oriented and is inherently different from the traditional organizational culture of schools.
- They already have experienced coping and working in an industry that functions in a very dynamic world and is constantly changing.
- Many of them have already managed and led change processes in their organizations and will be able to lead and implement this experience in the education system as well.

References

Aharoni D (2000) Cogito, ergo sum! Cognitive processes of students dealing with data structures. In: Haller S (ed) Proceedings of the 31st SIGCSE Technical Symposium on Computer Science Education, pp 26–30

Armoni M, Gal-Ezer J (2003) Non-determinism in computer science high-school curricula. FIE2003. http://fie-conference.org/fie2003/papers/1251.pdf. Accessed Nov 2014

Armoni M, Gal-Ezer J, Hazzan O (2006) Reductive thinking in computer science. Comput Sci Educ 16(4):281–301

Brandes O, Vilner T, Zur E (2010) Software design course for leading CS in-service teachers. Proceedings of ISSEP. Lecture Notes on Computer Science, Vol 5941, pp 49–60

CSTA (2013) Bugs in the system: computer science teacher certification in the U.S. http://csta.acm.org/ComputerScienceTeacherCertification/sub/CSTA_BugsInTheSystem.pdf. Accessed Aug 2014

Dubinsky Y, Hazzan O (2005) A framework for teaching software development methods. Comput Sci Educ 15(4):275–296

Gal-Ezer J, Harel D (1998) What (else) should computer science educators know? Commun ACM 41(9):77–84

Gal-Ezer J, Harel D (1999) Curriculum for a high school computer science curriculum. Comput Sci Edu 9(2):114–147

Gal-Ezer J, Zur E (2004) The efficiency of algorithms misconceptions. Comput Educ 42(3):215–226

Gal-Ezer J, Beeri C, Harel D, Yehudai A (1995) A high-school program in computer science. Comput 28(10):73–80

Haberman B, Lev E, Langly D (2003) Action research as a tool for promoting teacher awareness of students' conceptual understanding. ITiCSE 2003, pp 144–148

Hazzan O, Ragonis N (2014) STEM teaching as an additional profession for scientists and engineers: the case of computer science education, Proceedings of the 45th ACM Technical Symposium on Computer Science Education, Atlanta, GA, USA, pp 181–186

Hazzan O, Gal-Ezer J, Blum L (2008) A model for high school computer science education: the four key elements that make it!. Proceedings of the 39th ACM Technical Symposium on Computer Science Education, Portland, Oregon, USA, pp 281–285

Hazzan O, Gal-Ezer J, Ragonis N (2010) How to establish a Computer Science teacher preparation program at your university?—The ECSTPP Workshop. ACM, Inroads, pp 35–39

Israeli National Center for Computer Science Teachers (2002) "Machshava"-The Israeli National Center for High School Computer Science Teachers. Proceedings of the 7th SIGCSE Annual Conference on Innovation and Technology in Computer Science Education, Aarhus, Denmark, p 234

Kolikant Ben-DavidY, Pollack S (2004) Community-oriented pedagogy for in-service CS teacher training. ITiCSE 2004, pp 191–195

Lapidot T, Aharoni D (2008) On the frontier of computer science: Israeli summer seminars. Inroads SIGCSE Bull 40(4):72–74

Levy D (2000) Classification and discussion of recursive phenomena by computer science teachers. In: Robson R (ed) Proceedings of the International Conference on M/SET, San Diego, California

Lewin K (ed) (1948) Resolving social conflicts: Selected papers on group dynamics. Harper & Row, New York

Ragonis N, Haberman B (2003) A multi-level distance learning-based course for high-school computer science leading-teachers. ITiCSE, pp 224

Sherman-Kolker S (2009) Student *perceptions of human aspects of software engineering*, Master Thesis, Technion-Israel Institute of Technology

Shulman LS (1986) Those who understand: knowledge growth in teaching. Educ Teacher 15(2): 4–14

Tucker A, Deek F, Jones J, McCowan D, Stephenson C, Verno A (2003) A model curriculum for K-12 Computer Science. Final Report of the ACM K-12 Task Force Curriculum Committee. http://csta.acm.org/Curriculum/sub/K-12ModelCurr2ndEd.pdf. Accessed 20 Feb 2007

Epilogue

16

Abstract

This guide presents a comprehensive framework for the teaching of the Methods of Teaching Computer Science (MTCS) course as well as the teaching of additional other computer science and computer science education courses. We hope that it inspires the message that computer science learning and teaching processes can be fun, interactive, thought-provoking, and stimulating, and by delivering this message, learners' interest in computer science learning on all levels will be increased.

This guide presents a comprehensive framework for the teaching of the Methods of Teaching Computer Science (MTCS) course as well as the teaching of additional other computer science and computer science education courses. As has been mentioned in this guide, not all issues related to the teaching of these topics can be addressed in one guide. Indeed, we view this book as a guide that enables each computer science educator to further develop and adopt the material the guide presents for his or her individual needs. For example, the teaching of advanced computer science topics can be based on applying the principles presented in this guide, such as active learning, lab-based teaching, and the variations in teaching methods, types of questions, and tasks presented to the learners.

We do hope, however, that this guide does inspire the message that computer science learning and teaching processes can be fun, interactive, thought-provoking, and stimulating, and by delivering this message, learners' interest in computer science learning on all levels will be increased.

© Springer-Verlag London Limited 2014
O. Hazzan et al., *Guide to Teaching Computer Science*,
DOI 10.1007/978-1-4471-6630-6_16

Index

A

Abelson, H., 25, 38
Abhiram, 214
Abstract data types, 10, 265, 268
Abstraction, 10, 25, 44, 47, 48, 174
  level of, 39, 41, 42, 43, 216
Academia, 5, 36, 87, 247, 281
ACM Inroads magazine, 62
ACM K-12 Education Task Force, 2
ACM K–12 Task Force Curriculum Commit-
  tee, 270
ACM Task Force, 25
ACM Transactions on Computing Education
  (TOCE), 62
Action research, 278
Active learning, 8, 16, 17, 18, 143, 285
  in the MTCS course, 4, 17
Active-learning-based teaching methods, 9,
  105
Active-learning-based teaching model, 16, 19,
  20, 107, 160
Active-learning teaching approach, 246
Activity
  open-end, 19
Aharoni, D., 274, 275
Alan Turing, 30
Algorithm
  correctness, 77, 78, 171
  development, 77
  efficiently, 141
Algorithmic patterns, 9, 78, 79, 84, 85, 267
Algorithmic problems, 38, 57, 78, 84
Alice, 58, 199
Alternative conceptions, 13
  learners, 9, 59, 95, 267

Alternative consideration, 77
Ambler, A.L., 38
Analogies, 60, 78, 223, 264
Anderson, P., 49
Anderson, R., 16
Animation-based IDEs, 267
AP, 1, 138, 194
Armoni, M., 4, 57, 274, 275
Array
  boundaries, 212
  cell content, 213
  cell index, 212, 214, 216
  merge, 212
  of accumulators, 212
  of counters, 212
  of objects, 212, 213
  one-dimensional, 65, 70, 207, 211, 213
  scans, 212
  search, 212
  sort, 213, 217
Art, 26, 58, 77, 223
Arter, J., 200
Artificial inelegance, 30
Assessment
  formative, 188, 190, 196, 200
  peer, 189, 190, 210
  peer-assessment, 189, 190
  portfolio, 201, 202
  self-assessment, 189, 190
  summative, 188, 191, 196
Assignment, 4, 6, 13, 16, 19, 80, 215
Association for Computing Machinery
  (ACM), 30, 61
Astrachan, O., 84
Automata
  push-down, 268

© Springer-Verlag London Limited 2014
O. Hazzan et al., *Guide to Teaching Computer Science*,
DOI 10.1007/978-1-4471-6630-6

CPSIA information can be obtained
at www.ICGtesting.com
Printed in the USA
LVHW051936160519
618116LV00002B/41/P